TEACHING EFFECTIVELY WITH CHATGPT

First edition. July 1, 2024

ISBN

Ebook: 979-8-9908966-0-4

Paperback: 979-8-9908966-1-1

This book by Dan Levy and Angela Perez is a breath of fresh air. At a time when we all are swamped daily with abstract platitudes about the amazing potential of ChatGPT and A.I., both for teaching and for students' learning, this Levy and Perez book is unique in a specific way. It is chock full of concrete and specific examples – dozens of them - for improving teaching and learning when using ChatGPT. I heartily recommend this terrific book to any teacher.

Richard J. Light

Pforzheimer Professor of Teaching and Learning, Harvard University and author of "Making the Most of College" and "Becoming Great Universities"

In this original and highly readable book, Dan Levy and Angela Perez Albertos provide a range of suggestions for how professors and teachers can use Artificial Intelligence to improve their teaching. The book is structured using a helpful framework that covers the design and improvement of classes, the preparation of pre-class activities, instruction, assessment, and student support. Drawing on examples from their own teaching experience, and from the experiences of colleagues, Dan and Angela's engaging prose activates the reader's imagination about possible pedagogical experiments that will support more effective learning using machine learning algorithms trained on large language models such as ChatGPT.

Written in 2024, this book reflects the very early days of the incorporation of AI chatbots into teaching. It is not yet possible to know the direction of the transformation of teaching and learning that this tool will enable. It is certain, however, that such transformation will be informed by more experimentation of the sort this book invites the readers to undertake. For this reason, because it offers an invitation and the frameworks to include all readers in the exciting adventure of reinventing teaching and learning, this is a transformational book, which will have profound implications for the use of AI in education.

It should be read by anyone who wants to improve the effectiveness of their teaching. It will rekindle the curiosity and the joy of helping others learn.

Fernando M. Reimers

Ford Foundation Professor of Practice of International Education, Harvard Graduate School of Education

Dan does it again! His Zoom teaching book was a lifesaver in the early days of live online teaching. Now he and Angela Perez give us AI gold nuggets to help us create the classroom of the future. Their wisdom is based on their own experience plus that of colleagues around the world. Like its predecessor, this book will be a must-read for teachers at all levels, and especially in post-secondary and professional education. I can't wait to try some of these tricks and tools in both live-online and in-person classrooms. It pairs well with the Zoom book, and the companion site is a further gold mine.

Benjamin Gomes-Casseres

Peter A. Petri Professor of Business and Society, Brandeis Online Learning Lab at Brandeis University

May the hype cycles around generative AI and education finally now end, and the real, hard steps commence. Levy and Perez expertly walk readers through a reflective journey on how to actually use generative AI in ways that benefit student learning and help instructors be more effective. Rather than focusing on the bells and whistles of the tools, these cutting-edge instructors lay out clear recipes for success. We now have a cookbook rather than just a new stove and some pans. Let the next generation of teaching and learning commence!

Dustin Tingley

Professor of Government and Deputy Vice Provost for Advances in Learning, Harvard University

Feeling inundated by warnings, promises and information overload about Generative AI's impact on education? Take a walk with Dan Levy and Angela Pérez Albertos. Teaching Effectively with

ChatGPT, designed to help instructors "make sense of the chaos," is a welcome companion, written in hyperbole-free prose and chock full of pragmatic examples and strategies. As with Levy's earlier Teaching Effectively with Zoom, this guide is grounded in a solid pedagogical understanding of how students (and faculty) learn—and provides friendly advice that will make your teaching both more innovative and efficient. The lived experiences of the authors, and of the dozens of others they've consulted, provide detailed learning trajectories using ChatGPT with illuminating tips and debriefs. Knowing how quickly this landscape changes, Dan and Angela also have wisely and generously created a companion site for up-to-the-minute updates and resources. Don't "wait until the dust settles" (who knows when that will happen?); start exploring with this volume in hand, and you won't regret it.

Allison Pingree

Associate Director of Instructional Support & Development, Harvard Graduate School of Education

As executive educator and corporate board member, I gained instant insights from this book. Of course, many professors and executives are already aware of the watershed moment brought by generative AI. Everyone is anxious to be complemented rather than substituted by the new technology. But, for most, the issue lies neither in the awareness of the problem nor in the availability of tools. It is about overcoming inertia to adopt this technology. Precisely, Perez and Levy's book provides a comprehensive solution for adoption, offering a bridge with a straightforward learning curve. Their approach features clear, actionable steps that deliver immediate practical applications for both classrooms and boardrooms.

Rodrigo Wagner

Professor of Finance, Universidad Adolfo Ibañez Business School (Chile)

Companion site

Since ChatGPT and other AI technologies are evolving quickly, some of the prompts, instructions, and resources referenced in this book are located in the book's companion site, which can be updated more quickly than the book. The site is referenced multiple times throughout the book.

www.teachingeffectivelywithchatgpt.org

Angela:

To my family, mum, dad, Ana, and Javier; my partner, Laura; my teachers; and everyone else who has inspired me with their commitment to the pursuit of learning and the empowerment of others.

Dan:

To my wife Gaby, for her unwavering support during this journey and for enduring more conversations about AI than anyone should.

To my daughters, Dani and Alex, may you embrace a world transformed by AI with courage and wisdom.

To my parents, John and Licita, and my siblings, Vanessa and Ari, for shaping who I am as a teacher, learner, and human being.

Preface

In November 2022, ChatGPT stormed the world and soon became the fastest-growing consumer application in history, reaching 100 million users in its first two months.[1] Since then, a lot has been written about how we as educators should (or should not) incorporate this technology into our teaching, our students' learning, and our daily lives. The discussion at the beginning centered almost exclusively on the potential for students to use this technology to do the work for them, short-circuiting the process of learning. Some educational institutions, like the New York City Public Schools, temporarily banned the student use of ChatGPT for fear that it would lead to widespread cheating.[2]

Now that more time has passed, both the benefits and the limitations of this technology are starting to become clearer. Yes, students can use ChatGPT to cheat; but they can also use it to learn. In fact, it can be an amazing learning tool if used in the right way, akin to having a personal tutor with infinite patience available 24/7. Moreover, ChatGPT can be incredibly effective for instructors, assisting us in a variety of tasks ranging from brainstorming ideas for a class to summarizing student views on a particular topic, akin to having a teaching assistant with infinite patience available 24/7.

On the other hand, some challenges remain. Students can use ChatGPT in ways that hinder their learning to complete many, if not most, of the assignments we currently design. Furthermore, in most cases, it is hard for instructors (even when assisted with technology) to detect whether students used ChatGPT in a way that hindered their learning. From a practical perspective, it is becoming increasingly hard to know how much of the student work we assess was produced by the student and how much was produced by ChatGPT or a similar tool. ChatGPT is also known to produce content that sounds true, but is not, leading some to worry that the "hallucinations" it produces might lead our students (and ourselves) astray. And because ChatGPT is trained on vast amounts of data from the internet written by humans[3], its output reflects many of the biases that are present online. Finally, in an era where our students will be using AI tools in their work after they graduate, it is increasingly difficult for educators to discern which aspects of learning are foundational and critical and which may become less relevant, and therefore difficult to judge when the use of ChatGPT in assignments is or is not detrimental to learning.

We acknowledge both the benefits and risks associated with incorporating this technology into our teaching. But we also recognize that this technology is here to stay, that many of our students are already using it and that all our students will need to use it when they enter a labor market that will be profoundly transformed over the next decade. We are not the only ones who recognize that this is the case. In fact, a myriad of valuable advice on how to use ChatGPT in our teaching has appeared in blogs, articles, videos, etc. Which version of ChatGPT should I use? Are other large language models better than ChatGPT? Can it help me with this task? Why is it not good with this other task? For many of us, all these questions and well-meaning advice are overwhelming. Even paralyzing.

Additionally, the rapidly evolving nature of the technology means that keeping up-to-date with the latest advancements and understanding how to maximize their potential can be challenging. Some people may prefer to wait until the dust settles before committing time and resources to learn and adapt to new technologies. Our view is that the dust won't settle for a while and that it is best to start using the technology now and adapt to the coming changes rather than try to catch up years later.

This book is an attempt to help educators take a step back and make sense of the chaos. Our hope is to provide you with some useful pedagogic principles and practices that have served us and some of our colleagues well, and that we hope will guide you in your journey to developing your skills to leverage this technology for more effective teaching and learning. To help ensure that you can immediately apply the ideas in this book, we have included some concrete examples to illustrate how to implement some key practices in using ChatGPT. Whether you are new to ChatGPT (or generative AI more generally) or an experienced user, we hope you will find something of value in the book. If you are new to using ChatGPT in your teaching, our hope is that you will find a few ideas, try them out, and then experiment with additional ideas later in your journey. We also hope that this book will help you focus on what is important and provide a roadmap in your efforts to learn how to leverage this technology in your teaching. If you are an experienced user of ChatGPT in your teaching, we hope that some of the meta-advice here will be helpful and that you will pick up a few tips to improve your already well-developed practices.

Since the technology is evolving quickly, some of the instructions on how to implement these practices (including links to videos and ChatGPT prompts) are located in the book's companion site (www.teachingeffectivelywithchatgpt.org), which can be updated more quickly than the book. The companion site also contains links to additional resources, including readings, tutorials, and other AI tools.

The book is based on our own experience using ChatGPT, practices used by several colleagues, research-based principles of effective teaching and learning, and, perhaps just as importantly, interviews with dozens of students who are experimenting with using ChatGPT in their learning.

Who is this book for? We are both at a university, so the most natural audience for this book is people with teaching responsibilities in colleges and universities (faculty, instructors, instructional coaches, teaching assistants, etc.). However, after observing people of practically all ages use ChatGPT over the past year, we are convinced that K-12 teachers can also find value in this book.

Why ChatGPT? It's the predominant platform right now. It will make the advice in this book more concrete and grounded. Nevertheless, practically all the advice here applies to many other large language models (LLMs), such as Claude (from Anthropic), Gemini (from Google), and Llama (from Meta).

Who are we and why did we write this book? We are both passionate educators who believe that technology can play a positive role in improving teaching and learning. We also recognize the limits of technology and the very human nature of the process of education. Dan has been a faculty member at Harvard University for the past 20 years and wrote the book "Teaching Effectively with Zoom" which served as inspiration for this book. Angela worked as Chief of Staff to the CEO of the African Leadership Group (ALG) on several education initiatives and has taught or assisted in the teaching of several courses at Harvard. She also graduated in 2024 from a master's at Harvard and brings the very important perspective of the student to this book.

The book is organized as follows. **Part I** provides an introduction and a brief overview of the key overarching principles that we hope can guide you in incorporating AI into your teaching journey. **Part II** describes ways in which educators can use ChatGPT to teach more effectively, while **Part III** suggests ways in which our students can use ChatGPT to learn more effectively. Finally, **Part IV** describes some more advanced uses of the technology, concludes and suggests ways you can use what you learned to further develop your skills in using AI in your teaching.

If this book inspires you to do something you find interesting or exciting in your teaching that translates into your students engaging and learning more, please send us a note telling us about it. Nothing would be more rewarding for us.

Dan Levy and Angela Pérez Albertos
Cambridge, MA.
July 2024

Emails: dan_levy@harvard.edu and angelaperez@hks.harvard.edu

Acknowledgments

We would like to acknowledge the help of many people who have made this book possible. Four people deserve special mention. First, our colleague and Harvard Business School professor Mike Toffel, who gave us insightful feedback on several chapters of the book, helped refine our thinking and provided encouragement and support. Victoria Barnum superbly led the efforts to produce the companion site and to publish the book. Didi Milana played a key role in designing the companion site, and Erin Meade provided much needed editing support.

We would like to thank several individuals that shaped our views on AI in general, and AI and teaching in particular. They include José Antonio Bowen, Ben Brockman, Derek Bruff, Mark Fagan, John FitzGibbon, Seb Kaempf, Mae Klinger, Danny Liu, Lauren McHugh Olende, Dan Meyers, Ethan Mollick, Lilach Mollick, Allison Pingree, Sid Ravinutala, Chris Robert, Soroush Saghafian, Bruce Schneier, George Siemens, Al Stark, Mustafa Suleyman, Jason Tangen, Jim Waldo, C. Edward Watson, Mitch Weiss and Richard Zeckhauser. Dan is particularly grateful to Sharad Goel and Teddy Svoronos, with whom he taught a course titled "The Science and Implications of Generative AI" and who have been invaluable partners in leveraging AI at the Harvard Kennedy School. They taught him most of what he knows about AI. He is also grateful to participants in faculty workshops about leveraging AI for more effective teaching and learning at the Harvard Kennedy School, Harvard Graduate School of Education, Harvard School of Public Health, Harvard Medical School, SEK Education Group (Spain), Universidad Católica Boliviana (Bolivia) and Queensland Institute of Technology (Australia).

We are grateful to several people who gave us feedback on various chapters of the book. They include Mark Fagan, Mae Klinger, Eliana La Ferrara, Daniele Paserman and Teddy Svoronos. We would like to thank all the educators and students who allowed us to learn from their AI-related teaching practices, by making their materials available, speaking with us or submitting their examples and tips for leveraging AI effectively. Educators include Kimberly D. Acquaviva, Cynthia Alby, Molly Brady, Gregory Bruich, Fernando Díaz del Castillo, Raj Chetty, Vincent

Cho, Louis Deslauriers, David Dockterman, Lance Eaton, Bruce Ellis, Pary Fassihi, Mónica Flores Rojas, Nancy Gibbs, Sharad Goel, Gonzalo Jara, Evangelos Kassos, Robert Klitgaard, Tim Lindgren, Danny Liu, David J. Malan, Lilach Mollick, Ethan Mollick, Sean Norick, Cara Oneal-Radigan, Allison Pingree, Hong Qu, Fernando Reimers, Todd Rogers, Allison Shapira, Bruce Schneier, Teddy Svoronos, Jason Tangen, Mitchell Weiss, Matthew Wemyss and Bill Wisser. Students include Raghav Adlakha, Alexandra Arneri, Sofia Aron, Danielle Berger, Louis Guerin, Farid Hannan, Peter Huette, Sean Norick, Cara Oneal-Radigan, Shoroqu Othman, Hadar Sachs, Ryan Silber, and Isaac Davis Van Wert.

We are grateful to others who helped us with research, advice, insight, access to key information, and/or encouragement, including Bharat Anand, Erin Baumann, Aditya Bhayana, Josh Bookin, Matt Bunn, Suzanne Cooper, Anjani Datla, David Deming, Alejandro Díaz Garreta, Pinar Dogan, Erin Driver-Linn, Doug Elmendorf, Mark Fagan, Carol Finney, Maria Flanagan, Archon Fung, Ben Gomes-Casseres, Josh Goodman, Rema Hanna, Frank Hartmann, Ricardo Hausmann, Ron Heifetz, Dave Hirsh, Andrew Ho, Jim Honan, Kessely Hong, Doug Johnson, Asim Khwaja, Gary King, Jim Lang, Dick Light, Brian Mandell, Eric Mazur, Angelica Natera, Allison Pingree, Fernando Reimers, Dani Rodrik, Todd Rogers, Lori Rogers-Stokes, Miguel Angel Santos, Jeff Seglin, Allison Shapira, Kathryn Sikkink, Tyler Simko, Kristin Sullivan, Melissa Tarr, Dustin Tingley, Rodrigo Wagner, Marty West, Julie Wilson, Nick Wilson, Josh Yardley, David Zavaleta and Richard Zeckhauser.

Finally, apart from those students who contributed directly with examples that were featured in the book, many others provided us with valuable insights for the book. We are grateful to the students in several courses at Harvard (especially API-209, API-318 and DPI-681) who allowed us to experiment with the use of AI in teaching and gave us feedback on this experimentation, and to our teaching teams in these courses for being wonderful thought partners in our experimentation. Teaching team members include Amina Benzakour, Rohan Chandra, Madison Coots, Callan Corcoran, Shreya Dubey, Shira Gur Arieh, Shannie Lotan, Diletta Milana, Alvaro Andres Morales, Charlie Meynet, Sohaib Nasim and Dominic Valentino. We are also grateful to many others who participated in focus groups and informal sessions we conducted. Their perspectives, insights, and candor tremendously shaped our views and thinking.

Table of Contents

Part I - Introduction

Chapter #1 - Introduction

When we first heard about ChatGPT, we were both enthusiastic and skeptical. Enthusiastic because of its ability to instantaneously answer questions on all sorts of subjects in a way that often seemed indistinguishable from a knowledgeable human being. But skeptical about the claims that it would revolutionize education. As educators, we have seen countless tools come and go, each promising to transform our teaching and our students' learning. Yet many end up being more distracting than helpful. We see teaching and learning as an inherently human activity, making us naturally cautious about any technology that might be seen as a replacement for teachers.

Our experimentation with ChatGPT, combined with observations from its use by dozens of colleagues and students, has provided us with valuable insights into both its current potential and its limitations in complementing what educators do best. In this book, we share these experiences and insights with the hope that you can benefit from our journey. Our aim is to inspire and empower you to experiment with this technology, enhancing your teaching practice without being overwhelmed by it.

What this Book is About

This book is about helping you leverage generative AI (focusing on ChatGPT in particular) to teach more effectively and to help your students learn more effectively. Whether you are coming to this book reluctantly, enthusiastically, or mostly curious about using ChatGPT in your teaching, we hope this book will provide you with some guidance and ideas on how to leverage this technology effectively. There is so much advice out there that it can be overwhelming. And on top of it, AI is evolving at a neck-breaking speed. New AI tools and apps are created and marketed every day, focusing on functionalities from image creation and editing to task automation. Part of our goal in writing this book is to help you focus on what's important. You have limited time. You cannot spend countless hours aimlessly googling how to use ChatGPT, becoming a prompt engineer, or experimenting with the latest AI app. We tried to keep this reality in mind as we wrote this book.

This book is not a comprehensive catalog of everything you can do with ChatGPT in your teaching, but rather a guide for where to focus your attention and limited time. The overarching theme is that, at the end of the day, your goal is to help your students learn. Period. It is that simple. The technology is just the vehicle. You need to master some of the technology so you can focus on that goal. But you don't need to become a ChatGPT expert prompt engineer to do this well.

This book focuses on some useful pedagogic principles and practices that have served us and some of our colleagues well, and that will help you develop skills to use ChatGPT to teach more effectively. We use the term "develop" deliberately. We encourage you to think of this as a process where you will get better with practice. Just like you have done with other aspects of your job, time-saving benefits are unlikely to be realized immediately. As with onboarding a new assistant, it takes practice, getting to know each other, and working through the learning curve before productivity benefits materialize.

In sum, our goals in writing this book are to give you some ideas, empower you, and ultimately inspire you to leverage generative AI in your teaching. Our goal is not to overwhelm you or try to convert you into a ChatGPT expert.

What this Book is Not About

This book is not about using ChatGPT in general, although we hope you will become better at it for uses outside of the teaching realm. It is not about using the myriad of AI tools that are appearing constantly in the market either (although we cover some that may be useful in Chapter 11). While we recognize that at some point you might want to invest some time in those endeavors, we focus our attention on ChatGPT. There are three reasons for this. First, ChatGPT is very powerful in its own right, and trying to look beyond this tool as you start exploring generative AI tools is unlikely to be a good use of your time. Second, as ChatGPT evolves at such fast speed, there is a risk that other AI tools developed specifically for education may become outdated or simply disappear. Third, the principles you will learn using ChatGPT will likely apply to many other AI tools out there.

What this Book is Based On

The book is based on our learning from several sources: our own experience using ChatGPT to teach and learn more effectively, practices employed by other educators, research-based principles of effective teaching and learning, and, perhaps just as importantly, interviews with dozens of students on how this technology could be leveraged for more effective teaching and learning.

Although many of the approaches and practices recommended in the book are backed by research-based principles of effective teaching and learning, the focus will not be on the exposition of these principles and research findings. If you are interested in them, we recommend exploring the sources listed at the end of the book that pique your interest. The following books might also be of use to you if you are interested in learning more about some of the underlying research base of the learning sciences:

- *How People Learn: Brain, Mind, Experience, and School* (2000) by the National Research Council
- *How Learning Works: 7 Research-Based Principles for Smart Teaching* (2010) by Susan A. Ambrose, Michael W. Bridges, Michele DiPietro, Marsha C. Lovett, and Marie K. Norman
- *Make it Stick: The Science of Successful Learning* (2014) by Peter C. Brown, Henry L. Roediger, III, and Mark A. McDaniel
- *Multimedia Learning* (2009) by Richard E. Mayer

What Is the Approach of this Book?

This book is designed to be practical. We want you to gain some concrete ideas you can use in your courses. We have included actual prompts to illustrate how to implement some key practices in ChatGPT. We have also included ChatGPT's responses to these prompts, so you can see the tool's benefits and limitations. When the full ChatGPT response would occupy too much space and break your reading flow, we provide only an extract of the response instead.

We hope the principles about how to approach the use of AI in your teaching will be useful as the technology continues to evolve. But because we recognize that some of the instructions for doing certain things (e.g., uploading a file, doing data analysis, etc.) will inevitably be outdated as the technology evolves, we have created a companion site (www.teachingeffectivelywithchatgpt.org), which can be updated more quickly than the book. In this companion site, you will find all the prompts used in the book (so you can copy and paste them into your ChatGPT prompt box), instructions on how to do certain things on ChatGPT, information about AI tools that extend the capabilities of ChatGPT, and some video tutorials. We also want to invite you to use the companion site to share your own practices and ChatGPT dialogues, and to see what other readers have shared.

Should I Have a Paid Subscription to ChatGPT?

The answer depends on your personal circumstances. Previously, paid users had access to a substantially better model (ChatGPT-4) compared to free users (ChatGPT-3.5). However, as of this writing, OpenAI made available to both free and paid users its most powerful model to date, ChatGPT-4o, which has significantly reduced the benefits of a paid subscription. The paid subscription, ChatGPT Plus, costs US$20/month (as of this writing) and offers higher usage limits and the ability to create Custom GPTs (described in Chapter 10). While both of us have a paid subscription, we believe the gap between the free and paid versions has narrowed, so you might find the free version sufficient for your needs. Most of the advice in this book applies to both versions.

What do I Need to Know about ChatGPT?

ChatGPT is a powerful tool that belongs to a broader category of technologies known as artificial intelligence (AI). AI is a field of computer science that aims to create machines capable of mimicking human intelligence. Within AI, there's a specific area called generative AI. The term "generative" comes from the capability of these systems to generate new content, such as text, images, or music, rather than just analyzing or interpreting existing data. ChatGPT is an example of a generative AI model. It's powered by a type of AI known as a Large Language Model (LLM).

Sharad Goel, professor of Public Policy at Harvard Kennedy School and affiliate faculty member in Harvard's Computer Science Department, describes at a high level the building of LLMs as consisting of 3 steps:

1. Collect a large corpus of text generated by humans.
2. Based on this text, learn to predict the next word in any given sequence of words.
3. Fine-tune the language model to align with the desired behavior.

By repeating the second step over and over, ChatGPT can generate long strings of coherent text. This step of learning how to "predict the next word" is what people refer to as training the model.

For our purposes, there are two important implications of this three-step process. The first one is that ChatGPT was trained on a large corpus of text, not a large corpus of accurate, reliable, and unbiased text. While the third step of the process (fine-tuning) is designed partly to improve the accuracy of ChatGPT's output, this does not mean that the output will always be accurate and unbiased. This is a limitation of ChatGPT (and any other LLM) that we always have to keep in mind when using this technology.

The second implication is that the model is probabilistic when trying to predict the next word. Therefore, it will generally not give the same answers if you were to use the same prompt at two different times. To see this, let's enter the following prompt:

 Our first prompt

Please describe what good teaching looks like in one paragraph.

Here is the response we obtained when we typed the above prompt into ChatGPT.

 ChatGPT

Good teaching is a dynamic blend of passion, knowledge, and empathy that transforms learning into an engaging and impactful experience. It involves creating an inclusive and supportive environment where students feel valued and encouraged to participate actively. Effective teachers are adept at presenting complex material in a clear and accessible manner, using a variety of methods to cater to diverse learning styles. They foster critical thinking, stimulate curiosity, and inspire a love for learning by connecting theoretical concepts to real-world applications. Above all, good teaching is characterized by a genuine commitment to student success, demonstrated through continuous feedback, adaptability, and the nurturing of a growth mindset.

Hopefully, you have opened ChatGPT and typed this prompt yourself. If you compare the answer that ChatGPT gave you with the answer above, they will likely be similar but not exactly the same. In fact, if you went ahead and created a new chat with the exact same prompt for the second time, ChatGPT will not give you the exact same output as the first time!

Context Matters

We have tried to write this book to help educators in all sorts of institutions teach better with ChatGPT. In doing so, we have sought to recognize differences between institutional contexts. Some of you will have IT departments to consult with, teaching coaches to guide you, resources to create AI tools, and teaching assistants to help you. Others will just have a ChatGPT account and be pretty much on your own to figure things out for yourself. Moreover, some of you can expect all or most of your students to be able to access ChatGPT, whereas others will be teaching students who might not be able to access even the free version of ChatGPT. Finally, the type of course you teach might affect how useful ChatGPT can be for you. In general, ChatGPT will likely be more helpful for courses that rely on content that is widely available than for courses that are very specialized and reliant on knowledge that might not have been part of the large corpus of text that ChatGPT was trained on.

While we have tried to write the book in a way that, whatever your circumstances, you can get valuable ideas for your teaching, we recognize that context matters and that you will likely need to tweak and adapt some of these ideas to your institutional setting. Our goal is for you to find some - not all - of the ideas in this book useful to improve your teaching.

Security and Privacy

As mentioned above, Large Language Models such as ChatGPT are trained on a large corpus of text to enable them to generate responses that seem like they were generated by people. OpenAI, the company behind ChatGPT, also uses the information that we all type into their tool to fine-tune and improve the performance of their models. You should assume that anything you type or upload into ChatGPT is not private. As a result, a growing number of institutions are partnering with AI providers to use versions of their products that maintain the privacy of the information the users input, share, and expose to the AI tool. Unless your institution has made such arrangements and you are using the version of ChatGPT (or a similar AI tool) that has these protections, we suggest not exposing sensitive information to the system. For example, student grades should not be entered into ChatGPT. While most of the uses illustrated in this book do not involve what we would consider sensitive information, we suggest you keep an eye on this issue as you read the book.

How this Book was Written

This book was written by us (Angela and Dan) based on our ideas and informed by many people who have contributed to the debate on how to best use AI in education, including the many educators we feature in this book. Unsurprisingly, ChatGPT also contributed to the writing of this book. The first and most obvious way is that boxes with ChatGPT's responses to prompts are included throughout the book. ChatGPT also contributed in a variety of ways including summarizing key sources, turning some of our ideas into draft paragraphs, helping us incorporate the feedback we got from others and each other, brainstorming ideas, giving us feedback on the text we produced, and drafting generic text in response to prompts such as "write 2-3 sentences about the main benefits of one-minute papers." In sum, we stand on the shoulders of giants and ChatGPT!

How to Read This Book

The book is meant to be read in order, as there is a logical progression. However, the chapters are fairly self-contained, so if you are particularly interested in a chapter, you can skip ahead. Regardless of the order in which you read it, we suggest you keep nearby a device with ChatGPT installed (computer, tablet, or smartphone) and experiment right away (or at least at the end of each chapter) with the ideas from the book that you want to try out, and perhaps with some ideas not in the book that occurred to you while reading it. As indicated above, all the prompts are on the companion site (in case you want to copy and paste a prompt into your ChatGPT prompt box). You could also dictate a prompt to ChatGPT using its audio capabilities, which might be a particularly effective method if you only have your smartphone available while you are reading.

To save time when reading the book, we suggest you skim most of the boxes that contain ChatGPT's responses to the prompt. They are there to give you a sense of how ChatGPT responds but unless you are interested in the particular example, reading word by word might not be a good use of your time.

After you are done reading the book, you can review your book-inspired ChatGPT conversations (in ChatGPT's history) and see which ones you most want to pursue further. If you are new to using ChatGPT in teaching, you might want to re-read (or skim) some sections of this book after teaching for a few weeks, as you will likely have a new perspective and be ready to try some new approaches.

Getting Started with ChatGPT

The book will assume that you have some familiarity with the basic ChatGPT interface from having used it in the past. If you don't have a ChatGPT account or are unfamiliar with ChatGPT, please check introductory videos on the book's companion website (www.teachingeffectivelywithChatGPT.org).

Starting a New Chat

Figure 1.1 - ChatGPT's Basic Interface

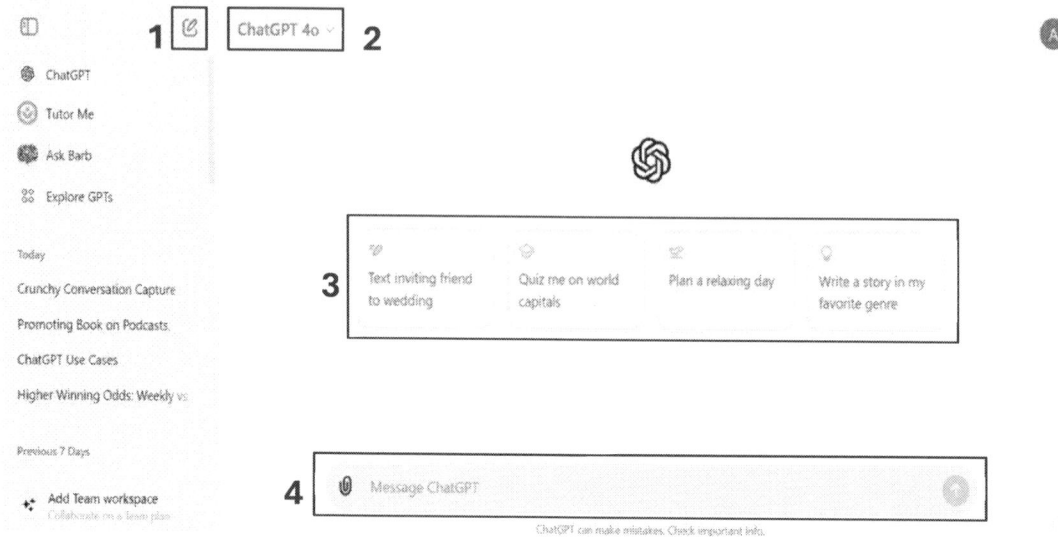

Here's a process you can follow to have a new conversation with ChatGPT:

1. Start a new Chat (or go back to a previous chat by searching in the left panel).
2. Check the version of ChatGPT in use - you can switch between ChatGPT 3.5, 4o, and 4 (if you have the premium version).
3. Select or be inspired by one of the conversation starters provided by ChatGPT.
4. If the conversation starters are not what you are looking for, begin your conversation with your own prompt.

Attaching Documents or Files

ChatGPT 4o can interact with files in various formats (text, images, audio and video). You may share documents like course syllabus, images, and other file types directly within the chat. However, given the lack of privacy highlighted above, you should be careful not to upload documents that are protected by copyright or contain sensitive or private information.

There are two ways to upload a file:

- Click on the "Upload" (paper clip) button, select the desired file from your device, Microsoft OneDrive or Google Drive, and it will be processed and integrated into the conversation;
- Drag the file directly from your directory into the chat.

Figure 1.2 - Uploading Files into ChatGPT through the "Upload" Button

Figure 1.3 - Uploading Files into ChatGPT by Dragging and Dropping from Your Directory

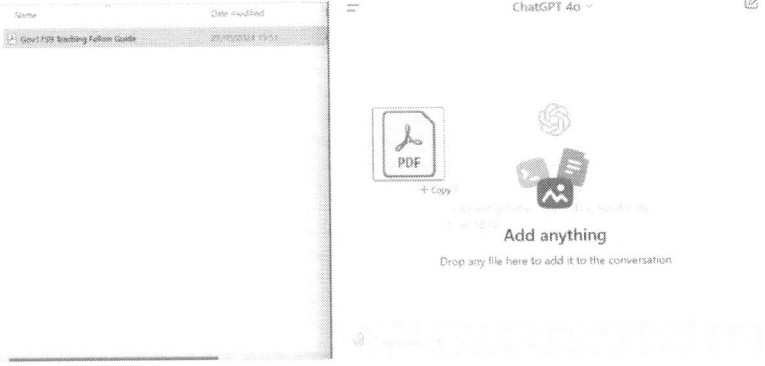

Custom Instructions

Custom Instructions are the initial set of instructions you provide to ChatGPT to guide any conversation you start with ChatGPT. They tailor ChatGPT's responses for more accurate and relevant output. Custom Instructions is a functionality that allows users to save time by training ChatGPT based on specific instructions, without the need to repeat these instructions in every new chat. As you will see later in the book, one of the key ways in which you can improve ChatGPT's performance is by providing extensive context about your role and the task to be performed. For example, you may provide ChatGPT with context about the course that you teach, the age of your students, the fields they are interested in, etc. However, because this can be repetitive and burdensome to do so each time you engage with ChatGPT, system prompts provide an important shortcut.

To insert Custom Instructions, you can click on your user logo, then click on "Customize ChatGPT". Please note that whatever you write in the Custom Instructions will apply to *all* of your conversations with ChatGPT, even the ones that are unrelated to your job as an educator. For this reason, we recommend that, if you decide to use them, you are fairly broad with what you write in the Custom Instructions.

Figure 1.4 - Adding Custom Instructions

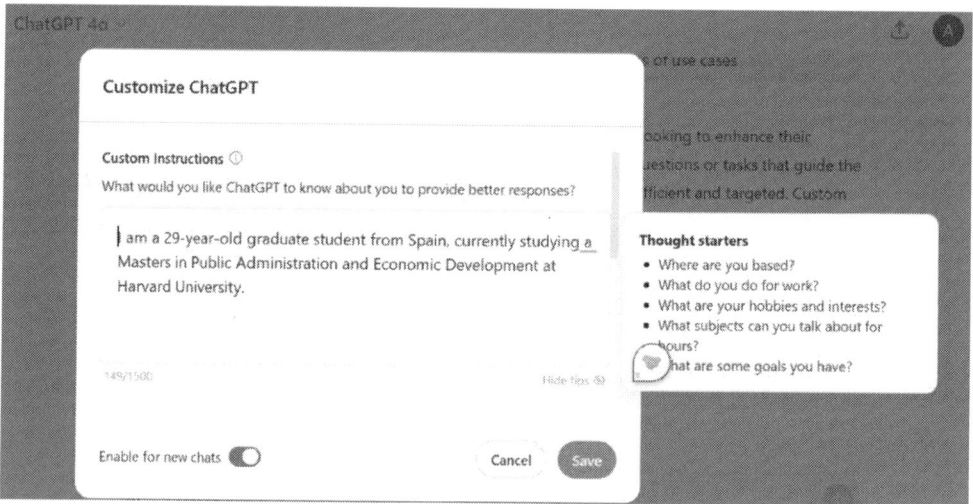

Custom GPTs

Custom GPTs are personalized versions of ChatGPT, tailored to meet specific needs or preferences. They are accessible through the "Explore GPTs" button you will find in the ChatGPT Sidebar. For example, the Logo Creator GPT helps users generate professional logo designs and app icons. Custom GPTs leverage the expertise of their creators to provide more specialized guidance. Popular custom GPTs for educators include TutorMe (a tutor developed by Khan Academy) and Consensus (which allows you to "chat" with the world's scientific literature). Chapter 10 will explore in-depth how to create and share custom GPTs with your students.

Disclaimer

ChatGPT is likely to evolve quickly. This book was written based on version 4o, the latest version available at the time of publication in July 2024. We have tried to write the book so that most of what is in it will remain true for some time. This is certainly true of the pedagogic advice. And we have relegated most of what we think will evolve to the companion website. But there are aspects of what's written here that will need to be updated in the future. So you can assume that every description of a ChatGPT feature or its default settings in this book should be preceded by "As of the time of writing, ..." We will be keeping close track of the changes to the platform that affect the advice in this book, and invite you to alert us to these changes in the feedback section of the companion website.

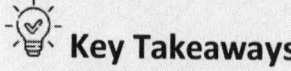 **Key Takeaways**

- Your main goal is to help your students learn. While it's important to understand some aspects of ChatGPT, you don't need to become an expert in prompt engineering to use it effectively.

- This book aims to provide you with practical ideas and strategies to integrate ChatGPT into your teaching, to empower you, and to ultimately inspire you to teach more effectively using ChatGPT.

- OpenAI uses the information from our conversations with ChatGPT to improve the performance of their models. Unless your institution has made arrangements to protect the privacy of what you share with ChatGPT (or any other AI tool), we suggest not exposing sensitive information to the system.

Chapter #2 - Guiding Principles

In this chapter, we describe three sets of key principles underlying the rest of the book: pedagogic principles, principles for using generative AI, and principles for effective prompting.

1. Key Pedagogic Principles

In planning to use AI tools like ChatGPT in your teaching, we encourage you to consider the pedagogic principles that underlie your teaching practices and how you can use ChatGPT to advance those principles. This section focuses on three key pedagogic principles:

Pedagogic Principle #1 – Be student-centered

Pedagogic Principle #2 – Plan for active learning

Pedagogic Principle #3 – Begin with the end in mind

Whether you agree with these principles or not, we want to make you aware of them because they underlie our advice for effectively using ChatGPT to advance teaching and learning.

Pedagogic Principle #1 – Be Student-Centered

Being student-centered means recognizing that teaching is not the same as learning. It requires teachers to focus on what students will be able to know or do as a result of their learning, rather than teachers focusing just on the topics they are covering. It's also about seeing teaching as a human enterprise, where your connection to your students and their connection to each other is vital to their learning process. Understanding your students' backgrounds, skills, and circumstances enables you to effectively adapt your approach to meet their learning needs.

ChatGPT can be a powerful ally in creating a student-centered learning environment. After you gather information about your students' interests, backgrounds, and learning preferences, ChatGPT can enable you to design more effective and engaging learning experiences. For example, tailoring pre-class activities and lessons to the specific needs of your students, you can create a more dynamic and responsive classroom environment (see Chapter 5). Alternatively, ChatGPT's ability to personalize learning can enable your students to learn at their own pace, pursue their curiosity by being in charge of asking questions, and follow their own learning pathway shaped by their background and interests (Chapter 8). This level of customization can increase their motivation to learn and be profoundly empowering for them.

Pedagogic Principle #2 – Plan for Active Learning

Effective teaching involves creating meaningful learning experiences where students play an active role. Research shows that students learn best when they actively engage with the material through processing, application, inquiry, and interaction. Planning for active learning means designing class sessions that encourage students to take charge of their learning and participate in activities that deepen their understanding.

ChatGPT can be a powerful tool in fostering active learning. It can help you design interactive activities – such as debates, simulations, discussion prompts, polling questions, and problem-solving scenarios – that require students to engage deeply with your course material (see Chapters 3, 4, 6, and 9). These activities can challenge students to think critically and collaborate with their peers, and ensure that your students are not passive recipients of knowledge but active participants in their learning journey.

Pedagogic Principle #3 – Begin with the End in Mind

This principle is about starting the design of any teaching activity by asking yourself what you want your students to achieve, rather than what you will "cover" in class. It is part of the "backward design framework", a widely used strategy proposed by Wiggins and McTighe (1998), which involves three specific steps, each of which can be supported by ChatGPT:

- Identify clear learning goals for your students. Chapter 4 illustrates how ChatGPT can assist in this process.
- Determine how you assess whether these goals have been met. Chapter 5 describes how to design pre-class work that allows you to assess where your students are before class with ChatGPT. Similarly, Chapter 7 addresses how to use ChatGPT to draft assignments that meaningfully test if your learning goals are being met using ChatGPT.
- Plan the instructional activities that will help students achieve these goals. Chapters 3, 4, and 6 include different ways in which you can leverage ChatGPT to design classroom activities.

Finally, there are a plethora of resources on how to implement this framework in your teaching. See our companion site for links to some of them.

2. Key Principles in Using AI for Teaching

In this section, we focus on key principles about the use of AI for teaching that underlie our approach in this book. These principles originate from our understanding of the learning sciences, our observation of fellow educators, and our own experience using AI tools to improve teaching and learning.

AI for Teaching Principle #1: Use AI to Augment Your Capabilities

Using AI for teaching is not a quest for having ChatGPT produce something better than what you would have produced. The question is not whether ChatGPT is better than you, but rather whether you and ChatGPT together can be better than you alone. However you define better (more learning, less time to prepare, more student engagement, etc.), we conjecture that you will find some areas in your teaching where you will conclude that you and ChatGPT together are better than you working alone. You can only discover these areas by experimentation. We hope this book will help you jump-start, accelerate, or inspire such experimentation.

Figure 2.1 - Key Test in Deciding Whether to Use Generative AI

Key test is **not** : AI > Me ?

Key test is: AI + Me > Me ?

AI for Teaching Principle #2: You are the Expert

As an educator, you are the expert. Given that AI can sometimes produce inaccurate information (what experts have dubbed "hallucinations"), your expertise is crucial in evaluating and enhancing the output generated by AI tools like ChatGPT. When ChatGPT provides information, your knowledge of your subject matter enables you to assess its accuracy and relevance. Even in areas where your expertise may not be strong, your critical thinking skills can allow you to determine the best ways to verify and evaluate the information. Furthermore, you understand your students better than anyone else. This allows you to adapt and shape ChatGPT's responses to meet the specific needs of your students. Finally, while ChatGPT offers general advice, you know what works best for you as a teacher in terms of your style, general approach, and preferences.

In sum, when using ChatGPT, remember that you are the expert and ChatGPT is your assistant. You are in the driver's seat; ChatGPT can be your co-pilot.

AI for Teaching Principle #3: Treat AI as Your Conversation Partner

Think of ChatGPT as a conversation partner. Imagine brainstorming birthday gift ideas for your partner with a friend. Your friend suggests a day at a spa, but you respond that this was your gift last year. You add that you generally like to provide gifts that are experiences, and you share more information about your partner's tastes and preferences. After some back-and-forth, your friend suggests you buy tickets to an upcoming concert of a singer your partner loves. While the first suggestion was not useful, the more context you gave your friend, the better able they were to help you. Similarly, you should expect to engage in a back-and-forth with ChatGPT. Provide feedback, ask follow-up questions, try different ways of asking the same question, and refine your queries, similar to how you would do with a conversation partner or assistant. Don't conclude from a response to a single prompt that ChatGPT cannot be helpful to achieve your goals; you will often need to engage in a conversation to get advice that's most useful to you.

AI for Teaching Principle #4: Experiment. Experiment. Experiment.

This book provides guidance for how to leverage ChatGPT to enhance your teaching, but the real key to mastering its use lies in practice. We hope the strategies and insights we offer are helpful, but to truly harness ChatGPT's potential, you need to experiment. Your specific needs and goals are unique; what is valuable to someone else may not be valuable to you, and vice versa. Try different approaches, see what works best to meet your specific goals, and adapt accordingly. Only through hands-on experimentation will you be able to fully unlock the benefits of ChatGPT for your teaching and your life more generally.

3. Key Principles in Prompting

Prompting is the process of giving ChatGPT specific instructions or questions to guide its responses. It's a bit like having a conversation where you steer the discussion by asking targeted questions or providing clear directions. Prompting is crucial because the quality of the output from ChatGPT greatly depends on the clarity and specificity of your inputs.

You might think that mastering prompting requires taking a specialized course (there are plenty out there; see links in companion site), but we don't think this is necessary. This section will give you the essential guidelines you need to get started using ChatGPT to enhance your teaching. By following these guidelines and experimenting on your own, you will quickly discover effective ways to use ChatGPT without feeling overwhelmed. While there is always more to learn about prompting, you don't need to know everything to start benefiting from ChatGPT. So after you read the guidance below, dive in, try out different approaches, and see what works best for you and your students.

Prompting Principle #1: Use A Simple Formula for Prompting

Prompt engineering typically relies on prompt formulas to improve the results users can obtain from ChatGPT. These formulas or frameworks can range from a list of concrete items to be included in the prompt to a full prompt draft. If you google "ChatGPT prompt formula," you will see thousands of results with titles such as "Master the Perfect ChatGPT Prompt Formula," "How to Craft the Perfect Prompt for ChatGPT," "The Only Prompt Formula You'll Ever Need for ChatGPT." Moreover, OpenAI provides six strategies for getting better results from prompting, and each of them with a set of tactics (see companion site for details).

All of this information can be overwhelming when you are just getting started. Our advice is simple: pick one formula and experiment. In the spirit of providing you with practical and concrete guidance, we describe below a formula that has served us well and that we use explicitly in several prompts in this book. The formula comes from our colleague Teddy Svoronos, a senior lecturer at the Harvard Kennedy School and teaching star, and was supplemented with guidance from Jeff Su, a consultant who has a YouTube channel with tips about AI and productivity at work more generally (see companion site for a nice video on his ChatGPT formula).

Table 2.1 - Simple Prompting Formula

Component	Description	Tips to Enhance ChatGPT Output
Task	What you want ChatGPT to do	* Be clear and specific

Instructions	How you want ChatGPT to do it	* Specify format of output (e.g., table, bullet points, etc.) * Specify tone of output (e.g., informal, authoritative, etc.) * Provide exemplars of the kind of output you are expecting * For complex tasks, ask it to go step-by-step * Ask ChatGPT to assume a persona
Context	What you want ChatGPT to know	* Specify info needed to do the task well (e.g., name of your course, background of your students, etc.) * Describe how the task fits in your broader project

The structure of "Task" + "Instructions" + "Context" is useful for thinking about all the key elements that may be required in any conversation with ChatGPT. The order in which you include these three elements does not seem to influence the results, at least in our experience. Also, you do not need to specify which part of your prompt corresponds to which element in the formula; the words "tasks," "instructions" and "context" do not need to appear in your prompt.

How much detail about your context is required to get helpful output depends on the task at hand. Think of it in a very similar way as if you were asking someone to do something for you. Imagine you are having some friends over for dinner and you are missing some items for the night. If you need ice, you may simply ask a friend to bring you "ice" with no more context, because you don't care about where they buy it, which shape, price, etc. But if you would like to buy a Manchego cheese to complete your cheese board and you know is only available in a gourmet store by the corner of your house, you will need to share all of these details with your friend. The same principle applies when you interact with ChatGPT.

Prompting Principle #2: When in Doubt, Ask ChatGPT What It Needs

If you are unsure about the appropriate amount of context to provide, consider asking ChatGPT for the criteria or details it needs to carry out the task properly. For example, imagine you want to write a series of tweets to promote an article you just wrote. Let's say you have never done this, so you don't know what details are relevant to carry out this task appropriately. Below is a similar example for us, the authors of this book. Since neither of us is very steeped in social media, we thought that asking ChatGPT for what context it needs would be helpful. As always, please note that whatever output ChatGPT produces (in this case, a series of tweets) is meant not so much to be the final product, but rather to provide ideas for the user to develop the final product.

 Dan and Angela

We are writing a book about teaching effectively with ChatGPT for educators. We would like your help in writing a series of tweets that we could use to promote the book. Please tell us what information you need to be able to write great tweets that are likely to get people to look at our book.

ChatGPT

To write effective tweets that promote your book about teaching with ChatGPT, I need the following information:

1. **Book Title and Subtitle**: The full title and any subtitles of the book.
2. **Key Selling Points**: The main benefits and unique aspects of your book. What makes it stand out from other educational resources?
3. **Target Audience**: Specific groups or demographics you aim to reach (e.g., higher education faculty, K-12 teachers, professional trainers).
4. **Notable Features**: Any particular features or chapters that are particularly engaging or innovative.
5. **Publication Details**: Release date, where the book can be purchased, any pre-order information, and formats available (e.g., print, ebook, audiobook).
6. **Authors' Credentials**: Brief background about you and any co-authors, highlighting your expertise in education and technology.
7. **Endorsements and Reviews**: Any notable endorsements, reviews, or quotes from early readers or influencers.
8. **Visuals**: Links to cover images, author photos, or any other visual elements you want to include in the tweets.
9. **Social Media Handles**: Any specific hashtags or social media handles you want to use or promote.

Once I have this information, I can craft a series of engaging tweets to effectively promote your book.

Note that a lot of what ChatGPT asked us is fairly obvious, but there might be a couple of items (such as #7 and #8) that we, as social media neophytes, might have forgotten to specify, and that might help produce a better output. In sum, ending some of your prompts with a phrase along the lines of "Ask me 5 questions that will help you provide me with the best possible response" might be helpful in general, especially when you are not too sure about what info ChatGPT needs.

Prompting Principle #3: When You Believe ChatGPT Made a Mistake, Tell It So

At times, you may think that an instruction is obvious and omit it from your prompt, only to see that ChatGPT is not returning the answer you expected. Other times, ChatGPT may be seemingly ignoring your instructions. Whatever the reason, if you believe ChatGPT has made an error, let it know so that it can offer more useful responses. For example, you may ask ChatGPT to summarize content from a web article you are interested in, but ChatGPT starts hallucinating and makes up a summary that is not based on the web article you shared. A simple nudge like "Your summary does not correspond to the article I shared. Make sure to browse the web and find the article that is linked" can help improve the answer. Another example: you may give ChatGPT a probability problem, and judge its first answer to be inaccurate. If you then specify "Please use probability theory to answer this question," the accuracy of the output will likely improve. If you think ChatGPT has made an error, point it out and see what it comes up with.

Prompting Principle #4: When Dealing with a Complex Task, Break It Down into Components

To help ChatGPT perform well, you might sometimes need to break the task into smaller parts. Generating intermediate reasoning steps helps guide ChatGPT's language model to a final answer and also helps you validate its reasoning process and output in each step. This is particularly helpful when handling complex tasks that require multi-step reasoning or problem-solving. This technique is called Chain-of-Thought (CoT) prompting[4]. See below a famous example for math problems (Wei et al., 2022).

Figure 2.2 - Using Chain-of-Thought Prompting

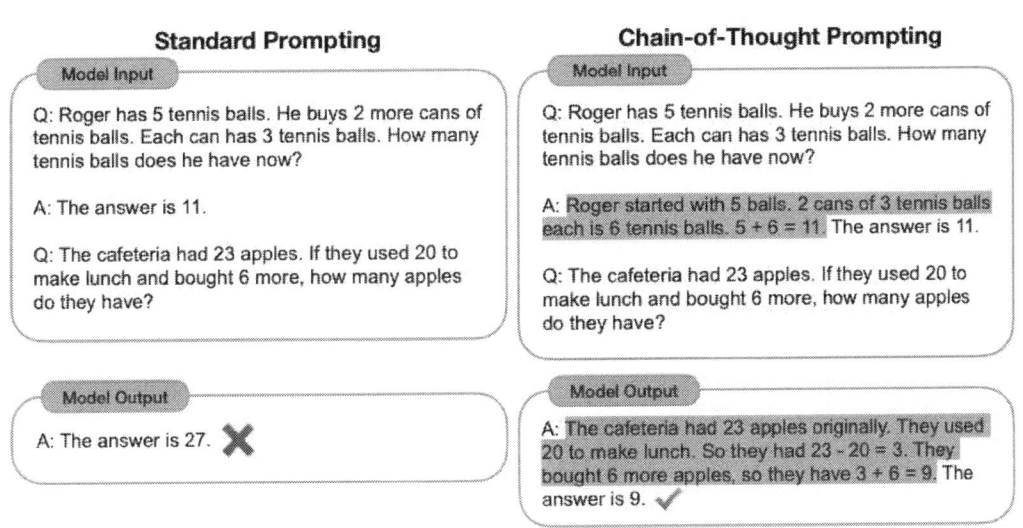

In the above example, the authors did an experiment with two arms: one using standard prompting (left column) and the other one breaking down the task into smaller steps (right column). For both arms of the experiment, the model was given an exemplar (the problem of Roger and the tennis balls). In the standard prompting case, the prompt included the correct answer (11), whereas in the Chain-of-Thought prompting case, the prompt also included the steps to solve this kind of problem. In the latter case, the authors of the paper were trying to "teach" ChatGPT (and the other LLMs they used) their process for solving this kind of problem. When ChatGPT attempted to solve the cafeteria problem, the model made a mistake in the standard prompting arm of the experiment, but "reasoned its way" to the correct answer in the CoT arm of the experiment by following the same problem-solving steps from the exemplar provided in the prompt.

While today's versions of ChatGPT can correctly solve the cafeteria problem even without exemplars (using the same step-by-step process outlined above), this general logic is still useful when asking ChatGPT to solve more complex problems. Breaking down a complex problem into components or steps can help get a better response from ChatGPT, just like it does for us human beings!

Prompting in Practice

While we encourage you to spend some time crafting the first prompts for your various uses of ChatGPT, keep in mind that if you forgot to write something on your first prompt, you can write it in your second prompt after seeing ChatGPT's output. Remember that you are meant to have a conversation to guide it (and you) to your desired destination.

Nonetheless, a first prompt can help set the context for more productive and efficient use of ChatGPT. To demonstrate this, we will compare the outputs of ChatGPT when using a simple prompt and when using one that complies with some of the guidance above.

 Chris (Simple Prompt)

I would like to learn about signal detection theory. Please explain this topic to me.

 ChatGPT

Signal Detection Theory (SDT) is a framework used to measure the ability to differentiate between information-bearing patterns (signals) and random patterns that distract from the information (noise). It originated in the fields of radar and communications but has since been applied to various areas including psychology, medicine, and decision-making.

Key Concepts of Signal Detection Theory

1. **Signal and Noise**:
     - **Signal**: The actual information or stimulus you are trying to detect.
     - **Noise**: Background information or stimuli that can interfere with the detection

of the signal.

2. **Hit, Miss, False Alarm, and Correct Rejection**:
   - **Hit**: Correctly identifying the presence of a signal.
   - **Miss**: Failing to identify the presence of a signal.
   - **False Alarm**: Incorrectly identifying a signal when there is none.
   - **Correct Rejection**: Correctly identifying that there is no signal.

3. **Sensitivity (d')**:
   - Sensitivity is a measure of how well one can distinguish between the presence and absence of a signal. It is influenced by the distance between the means of the signal and noise distributions. Higher sensitivity means better ability to discriminate signals from noise.

4. **Criterion (c)**:
   - The criterion is a threshold set by the observer for deciding whether a signal is present or not. It is influenced by factors such as expectations and the costs and benefits associated with hits and false alarms. A liberal criterion results in more hits and false alarms, while a conservative criterion results in fewer hits and false alarms.

5. **Receiver Operating Characteristic (ROC) Curve**:
   - An ROC curve plots the hit rate against the false alarm rate for different criteria. The area under the ROC curve (AUC) provides a single measure of the observer's performance, with larger areas indicating better performance.

Applications of Signal Detection Theory

- **Psychology**: Understanding perception and decision-making processes.
- **Medicine**: Evaluating diagnostic tests (e.g., sensitivity and specificity of medical screenings).
- **Communications**: Enhancing signal detection in the presence of noise.
- **Military**: Improving radar and sonar systems for better detection of objects.

Example Scenario

Imagine a radiologist examining X-ray images for signs of a tumor (signal) amid normal tissue (noise). SDT helps to quantify the radiologist's ability to detect the tumor accurately, balancing the need to minimize false alarms (e.g., unnecessary biopsies) and misses (e.g., undiagnosed tumors).

1. **Hit**: Correctly identifying an X-ray that contains a tumor.
2. **Miss**: Failing to detect a tumor that is present.
3. **False Alarm**: Incorrectly identifying a tumor in an X-ray that does not contain one.
4. **Correct Rejection**: Correctly identifying an X-ray without a tumor as normal.

By adjusting their decision criterion, the radiologist can balance the trade-off between detecting all possible tumors (sensitivity) and reducing unnecessary treatments (specificity).

Signal Detection Theory provides a robust framework for analyzing and improving decision-making processes across various fields by focusing on the interplay between signal detection and noise interference.

The above answer from ChatGPT is pretty good, but perhaps not so different from what we might obtain from a Wikipedia page or a Google search. Let's now see what happens with a more sophisticated prompt using the "Context + Task + Instructions" formula described above. One thing you might notice is that we (and our fictional character Chris) are very polite when having conversations with ChatGPT. We have no evidence that this is necessary or has any effect on ChatGPT's output, but we think it helps us develop better prompts by putting us in the state of mind of having a conversation with a colleague or assistant.

👤 Chris (More Sophisticated Prompt)

Context: I am a college student trying to learn about signal detection theory. I don't know much about this topic but my professor in a psych intro course asked us to learn about it for our next class. I am particularly interested in climate change, health, and education. I have taken a statistics course but don't remember much from it.

Task: Please explain Signal Detection Theory to me.

Instructions: Please use examples, and explain it to me as if I were a high school student. Please give me a short quiz at the end to make sure I understand. Make sure the explanation is no more than 3 paragraphs.

 ChatGPT

Signal Detection Theory (SDT) is a way to understand how we make decisions under uncertainty. Imagine you're trying to listen to your favorite song on the radio, but there's a lot of static noise. Sometimes you might think you hear your song when it's not really playing, and other times you might miss it when it is playing. SDT helps us analyze these situations by looking at how we detect signals (like the song) in the presence of noise (like the static).

In SDT, there are four possible outcomes:

1. **Hit:** You correctly detect the signal (you hear the song when it's playing).
2. **Miss:** You fail to detect the signal (the song is playing, but you don't hear it).
3. **False Alarm:** You think you detect the signal, but it's not there (you think you hear the song, but it's just static).
4. **Correct Rejection:** You correctly identify that there is no signal (you don't hear the song when it's not playing).

Let's apply this to health: Imagine you're a doctor trying to detect a disease from a test. The "signal" is the presence of the disease, and the "noise" is all the other things that might affect the test result. If you correctly identify a patient with the disease, that's a hit. If you miss a diseased patient, that's a miss. If you incorrectly diagnose a healthy patient with the disease, that's a false alarm. If you correctly identify a healthy patient as healthy, that's a correct rejection. SDT helps us understand and improve decision-making in such uncertain situations by balancing the chances of hits and false alarms.

31

As you can see, the answer is more tailored to the needs and preferences of Chris. In particular, note the initial practical application of signal detection theory was related to music. Specifically, how it helps us analyze why sometimes you might think you hear a song when it's not playing, and other times you might miss it when it is playing. This is a topic that someone in the high school/college stage of life can relate to. The second example was about health, which Chris had indicated as an area of interest.

Suppose that Chris also wants to explain signal detection theory to a friend but would like to use a different example than the ones related to music and health. Below is a possible prompt.

Chris

Now you are an expert in finding great and memorable examples of signal detection theory. Please provide me with 10 examples of how signal detection theory can be used in the real world. For each example, please explain what a "hit" and a "miss" mean. Please use an informal tone when responding. Use bullet points.

Here is an exemplar: Should you diagnose the patient with the disease? A hit means that you correctly identify a patient with the disease. A miss means you miss a diseased patient.

Please notice several elements in the prompt above, some of which were highlighted in the "Instructions" row of Table 2.1. First, Chris assigned a role or "persona" to ChatGPT (i.e., an expert in finding great and memorable examples of signal detection theory). This can sometimes be an efficient way of helping ChatGPT understand what kind of help you are looking for. Second, while Chris was looking for one example, asking for 10 examples meant a greater likelihood of finding one that Chris would appreciate. In general, err on the side of asking for more than what you need and then choose what you like from ChatGPT's output. This is one big advantage of ChatGPT over a human assistant: It doesn't cost more to deliver more output, and ChatGPT won't get tired or offended if you discard its output. Third, Chris specified the desired tone and format. Finally, Chris gave an example of what the output should look like. This can be an efficient way of indicating what you are looking for, especially when this is hard or cumbersome to do.

Below is ChatGPT's response. Even experts in signal detection theory would have difficulty in coming up with so many examples in such a short time.

 ChatGPT

Here are 10 real-world examples of how Signal Detection Theory can be used:

- **Detecting Spam Emails:** Should you mark an email as spam? A hit means you mark a spam email as spam. A miss means you leave a spam email in your inbox.
- **Taking an Umbrella:** Should you bring an umbrella when the weather forecast is uncertain? A hit means you bring an umbrella and it rains. A miss means you leave it at home and it rains.
- **Fire Alarm in a Building:** Should the fire alarm go off? A hit means the alarm goes off when there's a fire. A miss means there's a fire, but the alarm doesn't go off.
- **Finding a Parking Spot:** Should you take the parking spot? A hit means you take the spot and it's legal. A miss means you skip the spot, thinking it's illegal, but it was actually legal.
- **Deciding to Call a Friend:** Should you call your friend? A hit means you call your friend, and they are happy to hear from you. A miss means you don't call, but your friend was waiting for your call.
- **Checking for Typos in an Essay:** Should you correct a word? A hit means you find and

---

correct a typo. A miss means there's a typo you didn't notice.

- **Spotting Wildlife:** Should you take a photo of the animal? A hit means you take a photo, and it's a rare animal. A miss means you don't take a photo, but it was actually a rare animal.
- **Security Screening at the Airport:** Should you stop a passenger for further inspection? A hit means you stop a passenger, and they are carrying something dangerous. A miss means you let a passenger through, but they had something dangerous.
- **Dating App Matches:** Should you swipe right on a profile? A hit means you swipe right, and it's a good match. A miss means you don't swipe right, but they would have been a good match.
- **Detecting Lies:** Should you believe someone? A hit means you trust someone who is telling the truth. A miss means you don't believe someone, but they were actually telling the truth.

If you need more examples or further explanation, feel free to ask!

Notice that ChatGPT complied and gave Chris ten possible examples across a wide array of applications. Just like with any conversation, Chris could take this one in several directions. Below is one possible way to go. Chris continued by asking ChatGPT to format the outcomes in one table for future reference and to increase the difficulty of the quiz.

Chris

Great. The quiz you shared before is too easy. I can answer questions 1 and 2 by simply copying and pasting from above.

Please ask me questions that you haven't already answered in your explanation.

Also, can you please put the 4 possible outcomes in a table? I think this table could be a handy reference for me in the future. The first column has the name of the outcome, the second column has the generic description and the third column has the description in the umbrella example.

Explanation of Signal Detection Theory

Signal Detection Theory (SDT) helps us understand how we make decisions under uncertainty. Imagine you're a student trying to decide whether to bring an umbrella when the weather forecast is uncertain. Sometimes, you might bring an umbrella and it rains (hit), or you might leave it at home and it rains (miss). Other times, you might bring it and it doesn't rain (false alarm), or leave it at home and it doesn't rain (correct rejection). SDT helps analyze these situations by looking at how we detect signals (like rain) amidst noise (uncertainty in the forecast).

Outcome	Generic Description	Umbrella Example
Hit	Correctly detect the signal	You bring an umbrella and it rains
Miss	Fail to detect the signal	It rains, but you don't have the umbrella
False Alarm	Think you detect the signal, but it's not there	You bring an umbrella, but it doesn't rain
Correct Rejection	Correctly identify that there is no signal	You don't bring the umbrella, and it doesn't rain

Notice that ChatGPT did what Chris had suggested, including inserting a table. In general, ChatGPT can save you time by helping with formatting and editing tasks, such as converting text to tables, turning text into bullet points, sorting lists, etc. Chris can now answer the quiz questions and get some feedback from ChatGPT. We won't do this now, but in Part III of the book (Chapters 8-10), we explore ways in which ChatGPT can help our students learn in the kind of personalized way we just saw above and get feedback in the process.

Some Final Advice

We end this section with some quick practical tips for prompting ChatGPT.

1. **Start fresh when stuck:** If ChatGPT seems to be stuck in a loop or repeating itself, start a new conversation to reset the context. This helps in getting a fresh perspective. To do this, click on the pen and paper symbol at the top of the sidebar, or use the keyboard shortcut for your browser (command + shift + o for the Mac, and control + shift + o for Windows).

2. **Stop when unsatisfied:** If after you enter a prompt and ChatGPT starts generating a response that you know won't be helpful to you, you don't have to wait until it finishes composing the answer. You can stop it by clicking on the square symbol to the right of

the prompt box (this symbol is normally an up arrow but it turns into a square when ChatGPT is streaming text).

3. **Regenerate when unsatisfied:** If ChatGPT has finished generating a response and it does not meet your expectations, you can use the "Regenerate Response" feature to get an alternative answer. This can provide a different perspective or improved clarity. To do this, click on the small icon with two arrows at the end of every ChatGPT response).

4. **Insert paragraph breaks for long prompts:** When drafting a long or complex prompt, it might help to break it down into separate paragraphs. Sometimes when you press return (or Enter) the prompt gets sent to ChatGPT (instead of your cursor moving to the next line). Use Shift+Return on a Mac (or Shift+Enter on Windows) to insert line breaks in your prompt.

5. **Use keyboard shortcuts:** Chances are you will be spending more and more time using ChatGPT. Using keyboard shortcuts will likely help eliminate small frictions, which will improve your focus and efficiency. Below is a table with the shortcuts we found most helpful. Pick the one you think would be most useful to you, and commit to using it for a week. Once it becomes second nature, come back to this table for an additional one.

Table 2.2 - Main Keyboard Shortcuts When Using ChatGPT in Your Browser

Action	Mac Shortcut	Windows Shortcut
Open new chat	Cmd + Shift + o	Ctrl + Shift + o
Insert a line break in your prompt	Shift + Return	Shift + Enter
Send prompt to ChatGPT	Cmd + Return	Ctrl + Enter
Copy last response	Cmd + Shift + C	Ctrl + Shift + C
Toggle Sidebar	Cmd + Shift + S	Ctrl + Shift + S

Can you learn more about prompt engineering? Certainly. But just applying what you learned in this chapter will enable you to write great prompts to leverage a lot of ChatGPT's potential. So don't delay and start using what you learned right away!

 **Key takeaways**

- Three pedagogic principles to keep in mind: Be student-centered, plan for active learning and begin with the end in mind.
- Four principles for using AI to improve teaching: Use AI to augment your capabilities, you are the expert, treat AI as a conversation partner, and experiment with the technology.
- A possible formula for structuring your initial prompts for any endeavor: Context + Task + Instructions. Adapt it to your preferences and use it!
- If unsure about what ChatGPT needs from you, ask something along the lines of "Ask me 5 questions that will help you provide me with the best possible response".

Part II - Ways You Can Use ChatGPT

We are now ready to jump into action. This book is designed to offer practical guidance, featuring dozens of examples from real educators and our own experimentation. Beyond merely offering a list of prompts for you to use, we aim to bring the examples to life by walking you through the context that the educator faced, the purpose of the interaction, and the nuances of the information they provided. This second part of the book, which comprises Chapters 3-7, focuses on how you, as an educator, can use ChatGPT in your teaching. The next one (Part III) focuses on how your students can use ChatGPT.

To ignite your enthusiasm, we will begin by sharing the first use of ChatGPT that left us in awe – and was the moment we recognized ChatGPT's potential in education. It made us realize the benefits for educators in terms of efficiency and student learning. Inspired by the results of this experimentation, we joined forces to write this book and empower educators like you to leverage this technology in your teaching. Now, let's embark on this journey together.

This journey is based on an experience that Dan Levy, senior lecturer at the Harvard Kennedy School and one of the authors of this book, had in his initial experimentation with ChatGPT. In this example, Dan wanted to find a real-world example of decision-making to open his new course titled "Thinking Analytically in an Uncertain World." He wanted to use U.S. President John F. Kennedy's decisions during the 1962 Cuban Missile Crisis and wanted ChatGPT's help in building out the plan and resources for this class.

Dan Levy (Harvard Kennedy School)

Context: I am teaching a course titled "Thinking Analytically in an Uncertain World" for students who are doing a master's in public policy. Most of the students come with several years of work experience. The goal of the course is to help students think more analytically, which I hope will lead them to better understand the world around them, make smarter personal and professional decisions, achieve better outcomes, and ultimately lead a more fulfilling life.

Instructions: I would like to open the course with an example of decision-making that can anchor the course. The example is centered around the decision that President Kennedy and his administration made during the Cuban Missile Crisis. I am hoping the case can help illustrate some of the following concepts: decision-making under uncertainty, decision trees, risk, game theory, group decision-making, and how to avoid biases in group decision-making, etc.

Task: Can you please create an outline of a 60-minute class for me? I would like the class to be interactive. Also, please recommend resources I should read or watch to prepare for this class?

Notice the prompt has the three components mentioned in Chapter 2 (context, instructions, and task) but this time they are in a different order. That's OK. The structure we recommended is meant to remind you of elements that are useful in a prompt but not to be a straight jacket for you. Experiment with the structure that seems most helpful to you and assess the quality of the results you get with different prompting strategies. Below is ChatGPT's response, including a proposed outline and a proposal of resources to prepare for the class.

ChatGPT

Sure, here's a proposed outline for a 60-minute class centered on the Cuban Missile Crisis as a case study.

Outline

1. **Introduction** (10 minutes)
    - Course overview and objectives
    - Explanation of the day's topic: The Cuban Missile Crisis and its relevance to decision-making under uncertainty.
    - Brief background on the Cuban Missile Crisis.
2. **Case Study: Cuban Missile Crisis** (15 minutes)
    - Overview of the decisions faced by the Kennedy administration.
    - A quick video clip from the documentary "The Fog of War" on the Cuban Missile Crisis (clip chosen for relevance).
    - Discussion prompts: What decisions did Kennedy and his team face? What were the potential outcomes and consequences?
3. **Breakout Session** (10 minutes)
    - Divide students into small groups to map a decision tree of the Cuban Missile Crisis, identifying points of uncertainty and potential outcomes.
4. **Decision Trees and Risk** (10 minutes)
    - Presentation on decision trees, risk and their application to the case study.
    - Ask for a volunteer group to share their decision tree and discuss it.
5. **Game Theory and Group Decision-Making** (10 minutes)
    - Introduction of game theory concepts, applying it to the Cuban Missile Crisis (USSR vs USA).
    - Discuss the role of group decision-making during the crisis. How did the Kennedy administration work as a team? How did they avoid or succumb to group biases?
6. **Conclusion & Discussion** (5 minutes)
    - Summarize key learnings and insights from the case study and its application to

the concepts taught.
- Preview of next session's topics.
- Open up for Q&A or general reflections.

Preparation Resources:

1. **Books:**
   - "Thirteen Days: A Memoir of the Cuban Missile Crisis" by Robert F. Kennedy
   - "Essence of Decision: Explaining the Cuban Missile Crisis" by Graham T. Allison and Philip Zelikow
   - "Thinking, Fast and Slow" by Daniel Kahneman (for understanding biases)
2. **Videos:**
   - "The Fog of War: Eleven Lessons from the Life of Robert S. McNamara" (specifically the section on the Cuban Missile Crisis)
   - "Inside the Cold War: Soviet Union and the Cuban Missile Crisis - Documentary"
3. **Articles/Research papers:**
   - "The Cuban Missile Crisis, 1962: A Political Perspective After 40 Years" - James A. Nathan
   - "Game Theory" - Stanford Encyclopedia of Philosophy
   - "The Hidden Traps in Decision Making" - John S. Hammond, Ralph L. Keeney, and Howard Raiffa (Harvard Business Review)
4. **Online Courseware:**
   - MIT OpenCourseWare's "Game Theory and Political Theory"
   - Coursera's "Model Thinking" (for concepts on decision trees, game theory, and decision-making)

I hope this helps, and best of luck with your class! It sounds like a fantastic course.

A few things to note from ChatGPT's response. First, it seemed aware that this was the first class of the semester and suggested a section on course overview and objectives at the beginning of this class. This ability to draw implications of what seems like subtle cues is impressive to us. We suspect that if we had asked for similar advice from faculty members, not all of them would have been as situationally aware as ChatGPT seemed to be.

Second, notice that the phrase "I would like the class to be interactive" seems to have driven a few features in this class plan, including a breakout session, questions to discuss in class, and asking for a volunteer group to share their work. Third, notice that it listed a range of resources including books, videos, and articles. Some of the resources were particularly helpful to Dan. Among the books, it listed "Essence of Decision" which many experts consider one of the most important books on the subject. The book was written by Graham Allison, founding Dean of the modern Harvard Kennedy School, and a colleague of Dan. The documentary titled "The Fog of War" was highly relevant to understanding the thinking of a key actor (U.S. Defense Secretary Robert S. McNamara) during the Cuban Missile Crisis. Other resources, such as "Game Theory - Stanford Encyclopedia of Philosophy," were less relevant for this class. Notice that at this exploratory stage, identifying some relevant resources was the main goal, and ChatGPT was quite helpful.

Would Dan want to employ this class plan as-is? Unlikely. The goal was not to generate the perfect class plan with one prompt but rather to generate some ideas that you might consider useful for a class plan. Dan could have used some further prompting to have ChatGPT modify the class plan. For example, he could have asked ChatGPT for more details on how to organize the group work on section 3 or asked for a reallocation of time because he knows that conducting this kind of group work and debriefing the work is unlikely to take only 10 minutes in his class of 70 students. Given he was still trying to learn more about the topic, his follow-up prompt went in a different direction.

 Dan Levy (Harvard Kennedy School)

Thank you. Can you please tell me what you see as the main decisions that Kennedy faced and for each of them what options were considered?

ChatGPT

The Cuban Missile Crisis was a pivotal moment in world history when the potential for nuclear war was at its most tangible. Here were some of the key decisions and options that President Kennedy faced:

Decision 1: How to Respond to the Discovery of the Missiles

1. Options considered:
   - **Do nothing:** The U.S. could ignore the missiles, although this would risk the security of the nation and the region and set a dangerous precedent.
   - **Diplomatic negotiations:** Engage the USSR in diplomatic talks, potentially through the UN.
   - **Publicly call out the USSR and demand missile removal:** This would apply public pressure on the USSR but could escalate tensions further.
   - **Secret negotiations:** Approach USSR privately and negotiate missile removal, potentially in exchange for concessions from the U.S.

Decision 2: Choice of Military Action

2. Options considered:
   - **No military action:** Avoid escalating the situation by not responding with force.
   - **Air strikes:** Airstrikes to take out the missile sites in Cuba. However, this option risked escalating into a broader conflict or even a nuclear war.
   - **Invasion:** A full-scale invasion of Cuba to eliminate the missiles and potentially overthrow the Castro regime. This option held the greatest risk of escalating into a nuclear war.
   - **Naval blockade:** Enforce a naval blockade (or "quarantine" as it was officially called) to prevent further missile shipments from reaching Cuba. This was the chosen course of action, although it was risky as it could have provoked a military response from the Soviets.

3. Options considered:
   - **Publicly announce the crisis:** Publicly disclose the situation to the American people and the world. This option would allow the U.S. to control the narrative, but it would also escalate tensions and potentially provoke a response from the USSR.
   - **Keep the crisis confidential:** Maintain confidentiality to avoid causing panic and to buy time for negotiation. The risk of this option was the potential loss of public trust if the crisis became public knowledge.

These decisions showcase the immense complexity and uncertainty that Kennedy and his team faced. Each decision had several options, each with vastly different potential consequences. Making these decisions required careful analysis, group decision-making, managing various biases, understanding risk, and applying game theory principles.

Dan was impressed by the information provided on the decisions that Kennedy had to make and the speed at which ChatGPT made this information available. He also sensed that as he continued the process of preparing for this class, he would want to have this information handily available and therefore a more visual representation might be helpful. Hence, he asked ChatGPT to present the options in a table format.

Dan Levy (Harvard Kennedy School)

Thank you. Can you please provide them in a table format, and outline as additional columns the advantages and disadvantages that each decision presented?

ChatGPT

Certainly! Here's the reformatted table with the decision points aligned with their respective alternatives using merged cells for clarity:

Decision Point	Option Description	Advantages	Disadvantages
How to Respond to the Discovery of the Missiles	Do nothing	Avoid immediate conflict	Risks national security, sets a dangerous precedent
	Diplomatic negotiations	Could resolve the crisis peacefully, maintains global diplomatic channels	Slow, uncertain outcome, might not lead to missile removal
	Publicly call out the USSR and demand missile removal	Applies public pressure, demonstrates resolve	Could escalate tensions further, risking military confrontation
	Secret negotiations	May achieve discrete resolution, less public and global risk	If discovered, could appear weak or underhanded
Choice of Military Action	No military action	Avoids escalation, lowers risk of nuclear war	Perceived as weak, missiles remain in Cuba
	Air strikes	Could effectively remove missile threat	High risk of escalation, potential for Soviet and Cuban retaliation
	Invasion	Removes missiles, could overthrow Castro regime	Extremely high risk of nuclear war, international condemnation

	Naval blockade ("quarantine")	Stops additional missiles, shows firm stance without immediate aggression	Could provoke Soviet military response, international legal issues
How to Manage the Crisis and Communication	Publicly announce the crisis	Controls narrative, maintains public trust if managed well	Escalates tension, risks immediate Soviet response
	Keep the crisis confidential	Buys time for negotiation, avoids public panic	Risk of losing public trust if later discovered, high stakes if escalation occurs

This formatting helps to visually group each decision point with its options, making it easier to understand the connections and implications of each choice during the crisis.

Dan followed up with further prompts to learn more about how he could incorporate the Cuban Missile Crisis into the first class of his course, and for each of these, he obtained from ChatGPT similarly thorough answers (not displayed below for brevity's sake). These prompts included:

- Can you please tell me more about the most relevant group decision-making issues illustrated by the Cuban Missile Crisis?
- I have heard some commentary comparing the quality of decision-making in this instance (the Cuban Missile Crisis) vs in the Bay of Pigs. Can you please tell me a bit more about how the two decisions compared and how one might have influenced the other?

Since Dan was not an expert on the Cuban Missile Crisis, he asked his colleague Graham Allison who wrote a seminal book on this topic (listed above by ChatGPT) to review ChatGPT's output. Given that ChatGPT is not always accurate and that the stakes were fairly high, this was an important step to take. Graham was impressed and said, "This is the job I would expect from a top research assistant!"

Finally, Dan wanted to end class with a quick assessment. He asked ChatGPT to provide him with a couple of options on three of the topics in the class. Below is the dialogue. Dan would not want to use all of these questions in the limited time he had in class, but this is a menu to choose from. Besides, if he liked some of the questions he decided not to use in class, he could use them for other purposes, such as an assignment or exam.

 Dan Levy (Harvard Kennedy School)

You described above 3 key decisions that the Kennedy Administration faced and their choices, group biases they faced, and a comparison with the Bay of Pigs. Can you please suggest some "exit ticket" questions about the topics just mentioned that I can ask students to answer at the end of class that would give me a good sense of whether my learning goals were met? I am particularly interested in assessing their understanding of each of the following areas: (1) game theory, (2) group decision-making, and (3) how to avoid biases in group decision-making. Please create 3 questions for each of these areas (based on the class content): an easy one, a medium one, and a hard one.

Please make the easy one as a multiple-choice question and the other two as short essays. Remember these are students who are doing a master's in public policy. Let me know if you have any questions you would like to ask me so you can generate a great output.

Please note how Dan asked for three questions of varying difficulty for each topic. This helped him brainstorm more ideas to have options to use that would better adjust to the level of mastery that he wanted his students to have.

 ChatGPT

Certainly! Below are suggested "exit ticket" questions tailored to assess understanding of game theory, group decision-making, and biases in decision-making, particularly in the context of the Cuban Missile Crisis:

Game Theory

Easy (Multiple Choice):

During the Cuban Missile Crisis, the USA and USSR were engaged in a strategic interaction where both sides had to consider the other's potential responses. Which game theory concept does this scenario best illustrate?

    a.   Zero-sum game
    b.   Nash Equilibrium
    c.   Dominant Strategy
    d.   Tit-for-Tat Strategy

Medium (Short Essay):

Discuss how the concept of Nash Equilibrium might have been reflected in the decision-making process during the Cuban Missile Crisis. Provide specific examples from a situation where neither the USA nor the USSR could benefit by changing strategies unilaterally, assuming the other's strategies remained constant.

Hard (Short Essay):

Analyze the use of game theory in international relations with a focus on its application during the Cuban Missile Crisis. Evaluate the limitations of game theory in this context and discuss how these limitations could impact decision-making in current international crises.

Group Decision-Making

Easy (Multiple Choice):

What role did President Kennedy's Executive Committee (ExComm) primarily serve during the Cuban Missile Crisis?

    a.   Implementing military strategies
    b.   Providing diverse perspectives to aid in decision-making

c. Negotiating directly with the Soviet Union

d. Managing public relations and media

Medium (Short Essay):

Evaluate the effectiveness of the group decision-making process used by Kennedy's ExComm during the Cuban Missile Crisis. Consider aspects such as group composition, meeting structure, and the flow of information. How did these factors contribute to or hinder effective decision-making?

Hard (Short Essay):

Critique the decision-making process of the ExComm in the context of modern theories of organizational behavior and group dynamics. Suggest improvements or alternative strategies that could have been employed to enhance decision-making efficacy and outcome quality.

Avoiding Biases in Group Decision-Making
Easy (Multiple Choice):

Which type of cognitive bias, if not checked, could have most likely led to escalating commitments in Kennedy's decision-making process during the Cuban Missile Crisis?

a. Confirmation bias

b. Loss aversion

c. Sunk cost fallacy

d. Groupthink

Medium (Short Essay):

Reflect on the group biases that might have influenced the Kennedy administration's decision-making during the Bay of Pigs invasion. How did these experiences shape the approach to decision-making in the Cuban Missile Crisis?

Hard (Short Essay):

Propose a structured framework that Kennedy's administration could have used to better identify and mitigate biases during the Cuban Missile Crisis. Include specific methods or tools that could be integrated into the decision-making process to reduce the impact of these biases.

These questions are designed to assess both basic knowledge and deeper analytical skills related to the key concepts discussed in your class.

Dan reviewed the exit questions and chose the questions most appropriate for his learning goals and students. Wouldn't it be nice if these questions were also ready in a Google Form that allowed Dan to collect and analyze student answers more systematically? Creating the form manually would involve a tedious job of copying and pasting. It turned out that with a little bit of help and a spirit of adventure, Dan managed to do this with the help of ChatGPT. Here is the next stage of the dialogue.

Dan Levy (Harvard Kennedy School)

Thank you. I would like to do a survey for the students with these questions:

[...Questions omitted for brevity...]

Is there any way you could help me in the process of converting these questions into a Google Form format? Since the form is for students to fill in please avoid language about difficulty level or type of questions.

ChatGPT

I can guide you through the process of creating a Google Form based on the assessment, but I can't directly create or edit Google Forms myself. Here's a step-by-step guide on how to set up your assessment in Google Forms:

1. **Open Google Forms:**
   - Go to Google Forms.
   - If you're not already logged into your Google account, you will need to log in.
2. **Create a New Form:**
   - Click on the blank form or select a template from the gallery.
3. **Set Up the Form:**
   - Give your form a title, e.g., "Pre-Class Assessment on Bayes' Rule".
   - You can also add a description that explains what the form is about and any instructions you want the students to follow.
4. **Add Questions:**
   - For each question from the document:
     - Click on the "+" button on the right side to add a new question.
     - Choose the question type (e.g., multiple-choice for questions 1 and 2, paragraph for questions 3 and 5, and short answer for question 4 if you want them to enter the probability).
     - Copy and paste the question and the options from the document into the form.
     - Set the correct answer for multiple-choice questions by clicking on the "Answer key" if you wish to use the form for automatic grading.
5. **Customize Settings:**
   - You can adjust settings by clicking on the settings gear icon. Here, you can set whether respondents can edit their responses, whether they need to be logged into a Google account, etc.
6. **Send the Form:**
   - Once your form is ready, click on the "Send" button at the top right of the page.
   - You can send the form directly via email, or copy the link and distribute it as you see fit.
7. **View Responses:**
   - Responses to your form can be viewed by clicking on the "Responses" tab at the top of the form page.
   - You can also create a spreadsheet in Google Sheets to organize and analyze

> the responses more efficiently.
>
> This should help you set up the assessment in Google Form format. If you need further assistance or specific instructions on any of these steps, feel free to ask!

This step-by-step process of creating a Google Form felt overwhelming. Dan had heard from some tech-savvy colleagues that there was a way to automate the creation of Google documents through a tool called Google Apps Scripts.

Dan Levy (Harvard Kennedy School)

> Can you please write Google Apps Script code that allows me to take these 3 questions and convert them into a Google Form?

ChatGPT

> Absolutely! Here's a Google Apps Script to create a Google Form based on the three questions you've provided, formatted in a professional and clear manner for your students to fill out after class.

```javascript
function createSurveyForm() {
  // Create a new form and set its title and descrip
  var form = FormApp.create('Decision-Making Biases
    .setTitle('Decision-Making Biases Survey')
    .setDescription('Please answer the following que

  // Add the multiple choice question
  var biasesQuestion = form.addMultipleChoiceItem();
```

[...full code omitted for brevity...]

Instructions for Use:

a. **Go to Google Apps Script**: Open <u>Google Apps Script</u>.
b. **Create a New Project**: Click on "New Project", remove any placeholder script, and paste the script provided above.
c. **Save and Name Your Project**: Click the disk icon to save the script. Give your project a descriptive name.
d. **Run the Script**: Click the play button (▶) to execute the createSurveyForm function.
e. **Authorize the Script**: Follow the prompts to authorize the script, if required.
f. **Access the Form**: After running the script, check the logs (View > Logs) to find the URL of the newly created Google Form.

This setup ensures that the questions are presented in a clear and concise manner, appropriate for post-class assessment.

Dan clicked on "Copy Code", followed the instructions, and two minutes later had a Google Form that looked like this (only the first two questions displayed for brevity's sake).

Figure II.1 - Google Forms Quiz

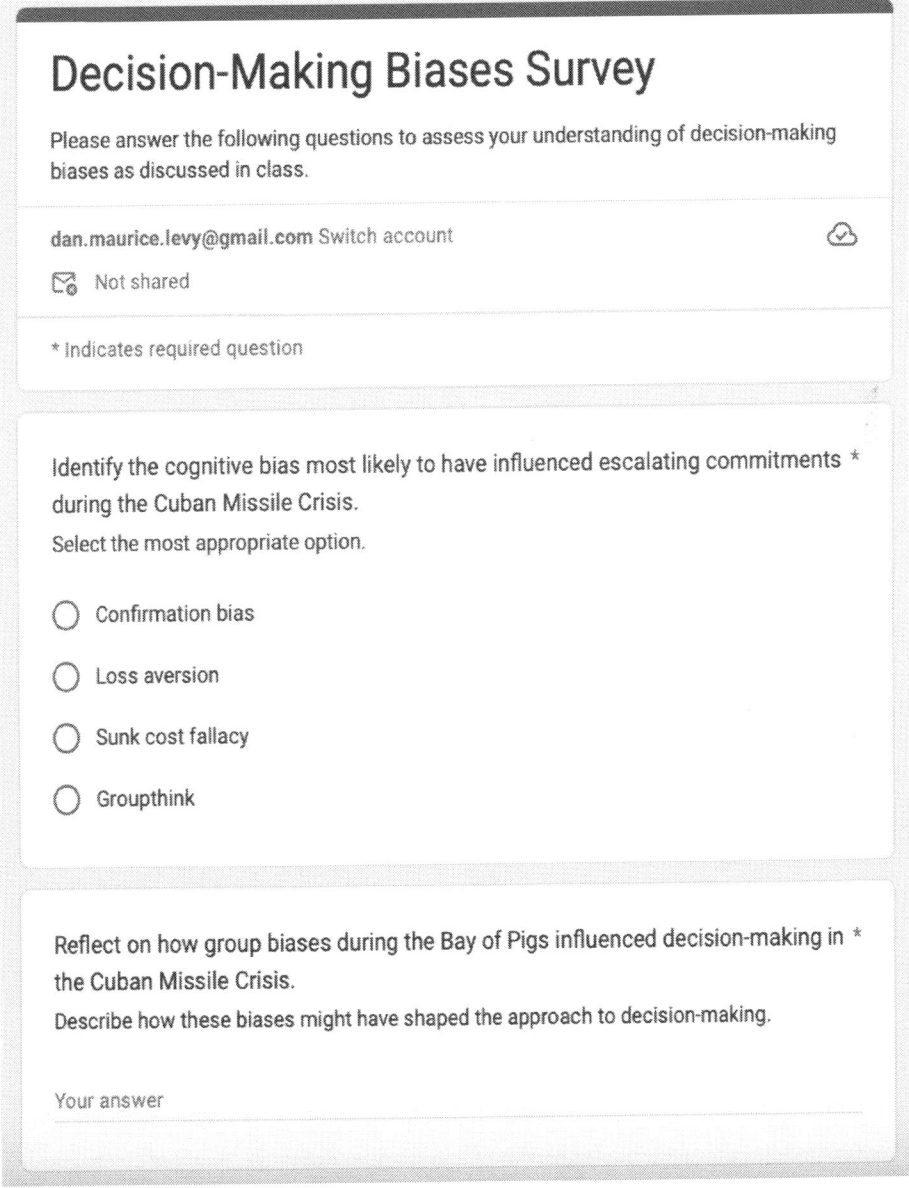

Dan was satisfied with the results but wanted to make some changes like changing the title, adding section titles, or adding question numbers at the beginning of each question. At this point, he could have decided to fix the form manually or to ask ChatGPT to implement the changes and produce a new Google Script. In this case, the changes were easy to make manually, but you could imagine the amount of time that could be saved if this was a longer form that required multiple, tedious changes.

Finally, as Dan was heading to class, he realized he had forgotten to create an easy way for his students to access the survey. He wanted to provide them with a QR code so they could quickly answer the survey on their phones. He could go into one of the services he normally uses to create QR codes. Instead, he typed the following into ChatGPT.

 Dan Levy (Harvard Kennedy School)

Can you please create a QR code for this survey link: https://forms.gle/otFGqrZo1vrBzLns7

 ChatGPT

Here is the QR code for your survey link:

Download QR Code. [...Link to QR code omitted...]

And just like that, Dan copies the QR code into his slides and heads to the classroom!

Figure II.2 - QR Code Generated by ChatGPT for End-of-class Survey

To conclude, a few things to highlight from this example:

- This was a long conversation between Dan and ChatGPT, where ChatGPT acted at times as a research assistant (e.g., when listing the decisions that President Kennedy faced), at times as a teaching assistant/coach (e.g., when suggesting an outline for a 60-minute class), and at times as and as an IT assistant (e.g., when writing code to create the Google survey, and to generate the QR code). These assistants were available 24/7.

- The conversation above took place over several days. When coming back to the chat, Dan could remind himself of where he left off by simply reading the chat history. And ChatGPT needed no reminder of where it was.

- In the process, Dan had to use his expertise and judgment to assess ChatGPT's output and decide which parts to use, which parts to discard, and which parts to investigate further. In this particular case, since he lacked expertise on the Cuban Missile Crisis, he consulted a trusted expert colleague to assess the accuracy of ChatGPT's output. Access to such experts may not always be available to you, in which case finding other ways to fact-check the information would be essential.

- On some occasions, ChatGPT did not do what Dan wanted. For example, as you were able to see above, the initial response to Dan's request to create a Google form gave instructions that were not very helpful for Dan. So he had to prompt again. Dan had a similar experience (not shown above) when asking ChatGPT to create the QR code. In this case, he opened a new chat, asked again, and got the QR code.

In sum, we hope this example illustrated how ChatGPT can assist you in creating better learning experiences for your students and save time in the process (hence the sub-title of this book: "A practical guide to creating better learning experiences for your students in less time"). In the chapters ahead, we will dive more deeply into how ChatGPT can help you in different aspects of your teaching: improving a class session (Chapter 3), designing a new class session (Chapter 4), designing pre-class work for your students (Chapter 5), using ChatGPT inside your classroom (Chapter 6) and designing assessments (Chapter 7). Read ahead!

Chapter #3 - Improving an Existing Class

In this chapter, we assume that you would like to improve a class session that you have taught in the past. Perhaps you are happy with the way the class went but would like to improve a specific aspect of this upcoming version of the class, such as streamlining your slides, making the class more engaging for your students, or coming up with better examples to illustrate a concept or idea. The chapter assumes you want to tweak some of what you did the last time you taught the class rather than recreate the class from scratch. If you are interested in the latter or more generally in using ChatGPT to design a new learning experience, please see "Chapter 4 - Preparing for a new class session." This chapter, like all others in this part of the book, builds on the pedagogic principles outlined in Chapter 2.

Below are some examples of the kinds of things you can do to improve a class.

Table 3.1 - Ways of Improving an Existing Class

Way	How ChatGPT can help
1 - Improve your slides	* Assess the strengths and weaknesses of your existing slides, and suggest changes * Suggest ideas for new slides
2 - Update your class activities	* Brainstorm ideas for new activities * Produce supporting material, e.g. slides, handouts
3 - Generate explanations, examples, or analogies	* Brainstorm different explanations of a concept/idea * Adapt explanations to the background of your students
4 - Improve your class plan	* Brainstorm ways of improving your class session

	relative to what you did last time
5 - Summarize student feedback	* Summarize key themes in course evaluations * Suggest how to improve your teaching
6 - Get feedback on your teaching	* Assess strengths and weaknesses of a class you taught * Give you suggestions for improvement on a class you taught

1. Improve Your Slides

Many instructors use slides to present material they want to teach and to assist them in conducting discussions in their classrooms. A good design and use of slides can be part of an effective teaching strategy. ChatGPT can help provide feedback on your slides and the way you use them in the classroom. Below is one example from Dan. In this example, he uploaded to ChatGPT a PowerPoint file that he had used in an executive education program in the previous year and asked for advice to implement for this year's program. Quick reminder: to upload the file, you can simply drag it into ChatGPT's dialogue box or click on the paperclip item on the left side of that box.

 Dan Levy (Harvard Kennedy School)

Context: I am a faculty member at a university looking for ways to improve my teaching. Attached you will find a PowerPoint presentation for a session I gave last year in one of our executive education programs. The participants are professionals from the for-profit, nonprofit and public sectors who are interested in gaining insights they can use in their work. They come from all over the world. I am looking to improve the session and would like your help.

> **Task**: I would like the PowerPoint presentation to be as engaging as possible using the principles of effective slides. I would like to conduct the session using pedagogic principles of effective teaching and learning. Please give me advice on how to improve my slides. Later I will ask you for advice on how to improve the way I conduct the session, but for now, let's focus on the slides.
>
> **Instructions**: The advice should be specific and easy to implement. For example, on slide 11, break the text into shorter sentences or bullet points. Or for slide 13, include a simple example of a decision tree diagram. Please give the advice slide by slide. My session is tomorrow, and I don't have time for drastic changes.

Notice that for the instructions, Dan gave a few exemplars to try to ensure that he would get the type of feedback that would be most helpful to him. Also note that by writing "My session is tomorrow, and I don't have time for drastic changes" he is also shaping the kind of advice he is seeking. ChatGPT's response is rather long, and as indicated earlier in the book, we encourage you to skim ChatGPT's responses just to get a sense of the type of output you can expect.

 ChatGPT

Let's go through your PowerPoint presentation slide by slide and provide specific, easy-to-implement advice to make it more engaging and effective.

Slide 1: Title Slide
Suggestion: Add a relevant image or background that represents the session's theme to make the title slide more visually appealing.

Slide 2: Goal of This Session
Suggestion: Use bullet points for the key elements (Intuition, Frameworks, Evidence) instead of a single block of text. This makes it easier to read.

Slide 3: How Important is Your Intuition in Making Decisions?
Suggestion: Use a larger font size for the question and the QR code link to ensure they stand

61

out. Include a brief instruction like "Participate using the QR code or the link."

Slide 4: Birthday Puzzle
Suggestion: Break the information into two slides. One for the puzzle setup and another for the audience's guesses. Use visuals (e.g., icons or simple diagrams) to illustrate the puzzle.

Slide 5: Answer to Birthday Puzzle
Suggestion: Present the answer with a large, bold font. Include a brief explanation or visual representation of how the probability is calculated.

Slide 6: Meta Lessons
Suggestion: Use a mix of text and visuals (e.g., a picture of Kahneman). Reduce text by summarizing key points in bullet form.

Slide 7: Behavioral Biases - Examples
Suggestion: Use icons or illustrations for each bias to make them more memorable. Keep the descriptions concise.

Slide 8: Behavioral Biases – Some References
Suggestion: Create a single slide with a list of recommended readings. Include book covers to make it visually engaging.

Slide 9: Chances that at Least Two People Share a Birthday
Suggestion: Use a bar graph to illustrate the probabilities. Visual data representation can make the information more accessible.

Slide 10: Is Intuition Irrelevant?
Suggestion: Highlight the main point in bold and use a relevant image or visual metaphor (e.g., a plane in a cloud to represent the 'graveyard spiral').

Slide 11: Decision Analysis Frameworks
Suggestion: Introduce a decision tree diagram to visually represent the framework. Keep the text minimal and focused.

Slide 12: Key Steps in Most Decision Analysis Frameworks

Suggestion: Use bullet points and a flowchart to represent the steps. Visual aids help in understanding the process better.

Slide 13: Decision Trees

Suggestion: Include a simple decision tree example. Visuals help in comprehending abstract concepts.

Slide 14: Examples of Best Decisions

Suggestion: Use a consistent format for each example. Include photos of the individuals if possible, and keep the text short and to the point.

Slide 15: Quality of Decision vs. Quality of Outcome

Suggestion: Use a 2x2 matrix to illustrate the concept. Visual aids can clarify complex ideas.

Slide 16: What Drives the Outcome of a Decision?

Suggestion: Use icons or images to represent 'Skill' and 'Luck'. Keep the text brief and engaging.

Slide 17: Use of Evidence

Suggestion: Add a relevant image (e.g., a microscope or research icon) to emphasize the importance of evidence.

Slide 18: Malaria Case Study

Suggestion: Use a mix of images and text. A photo of bed nets being used and a brief summary of the case study can make it more engaging.

Slide 19: What is the Optimal Price for Bed Nets?

Suggestion: Use a bar graph to compare the different prices and their impacts. Visuals make data easier to digest.

Slide 20: More Recent Evidence

Suggestion: Summarize key findings with bullet points and include visual aids like charts or graphs to support the data.

Slide 21: Impact of Textbooks in Kenya
Suggestion: Use a pie chart or bar graph to show the impact. Visual representation of data is more engaging.

Slide 22: Reflections
Suggestion: Use a larger font for the instructions. Include a relevant image (e.g., a thought bubble) to encourage reflection.

Slide 23: Thank You Slide
Suggestion: Include a photo of yourself and contact information to personalize the slide. A warm, personal touch can make a big difference.

Backup Slides
Suggestion: Ensure they are well-organized and concise. Use visuals where possible to support the text.

By implementing these suggestions, you can make your PowerPoint presentation more engaging and effective, enhancing the learning experience for your participants. If you have any specific slides you'd like further details on, please let me know!

Much of ChatGPT's advice is focused on making the information more visually attractive and accessible. While this may seem trivial, small changes in how information is presented can positively affect student engagement and ultimately their learning. Moreover, if Dan was looking for a different type of advice (say, on incorporating more active learning in the class), he could have been more specific in his initial prompt or any follow-up prompt (see next section for an example). Finally, not all the advice above is useful to Dan. Yet, there are a couple of suggestions that Dan wants to implement and can ask ChatGPT for further assistance.

 Dan Levy (Harvard Kennedy School)

> On slide 8, you mentioned "Create a single slide with a list of recommended readings. Include book covers to make it visually engaging." Can you please give me links to the book covers so I can easily copy and paste them into my slides? Please make sure the links work. Take your time.

Dan could have conducted four web searches and found the images, but with this simple prompt, he was able to get links to all four images at once. Notice that because Dan has had occasions where the links that ChatGPT gives don't work, he made a special note to ChatGPT at the end to check that the links would work. ChatGPT obliged and produced links to the 4 books listed in that slide.

 ChatGPT

Here are the links to the book covers with clickable URLs:

1. "Thinking, Fast and Slow" by Daniel Kahneman (2011) *[...Link to image omitted...]*
2. "Predictably Irrational" by Dan Ariely (2008) *[...Link to image omitted...]*
3. "Nudge" by Richard H. Thaler and Cass R. Sunstein (2008) *[...Link to image omitted...]*
4. "Misbehaving: The Making of Behavioral Economics" by Richard H. Thaler (2015) *[...Link to image omitted...]*

You can click on these links to access and download the images from Amazon.

Dan could further inquire about other suggestions he wanted to act on. Below is one of the several ways to get further assistance from ChatGPT.

 Dan Levy (Harvard Kennedy School)

For slide 16, you suggested "Use icons or images to represent 'Skill' and 'Luck'. Keep the text brief and engaging." Can you please produce images for skill and luck that you think would work well with this slide? Make sure the images match (or at least are not too inconsistent) with the slide's theme.

ChatGPT produced two images along with the text below. Please note that if Dan did not like these images, he could ask ChatGPT to produce other images or to find images on the web that he could consider. Even if he did not like the particular images but liked the idea behind them (e.g., the gear and wrench to represent the word skill), he could still use the idea to search for similar images on the web or produce images on his own (see below). Using AI for slide production and editing is an area that is rapidly changing. Please see the section on slide production in Chapter 4 for more details, and our companion site for some resources and tutorials.

 ChatGPT

Here are two professional icons representing "Skill" and "Luck." You can choose the one that fits best with your presentation theme.

- The first icon features a gear and wrench, symbolizing expertise and ability for "Skill."
- The second icon features a four-leaf clover, symbolizing good fortune for "Luck."

You can incorporate these icons into your slide to make the concepts visually engaging.

Image 1	Image 2

Finally, you might be wondering at this point whether or how to cite ChatGPT contributions (such as the images above or any text) to the work that you do. This is a field that is rapidly evolving with different institutions applying different norms. The companion site links to an article that highlights some of the issues in this debate. We suspect that given the way that the technology is evolving, it will likely become increasingly difficult to clearly distinguish between your own output and AI-generated content.

2. Update Your Class Activities

As described in Chapter 2, an important pedagogical principle underlying this book is to plan for active learning. This section assumes that you have a desire to do so for your upcoming class but don't want to completely change what you did the last time you taught this same class session. So you may be looking to insert a short activity or tweak an activity rather than design activities that could take a long time to do in the classroom (such as a simulation or a case discussion). If you are interested in the latter, please see the next chapter.

We will continue with the example from our last section, where Dan asked for feedback on a set of slides he used in an executive education program.

 Dan Levy (Harvard Kennedy School)

I am now looking for ways to incorporate active learning into this class. Since the class is tomorrow and I don't have time to overhaul the class, I would like you to look at the slides again and suggest places in the class where I could insert an activity that would make students actively engage in their learning.

These activities could include answering a poll for an important question that can help me check for understanding or help them grapple with an important idea in the class, discuss a topic or do an exercise in small groups, or have a class-wide discussion on a question important to the class goals.

Each of your suggestions should take no more than 5-10 minutes to implement in the classroom. Please begin by asking me what else you need to know to be able to give me useful advice.

Notice that since Dan was not sure that ChatGPT had all the information needed to produce useful advice, he simply asked ChatGPT what kind of information it needed by asking: "Please begin by asking me what else you need to know to be able to give me useful advice." Below are the questions that ChatGPT asked Dan as well as his responses.

 ChatGPT

To give you the most useful advice, I need to know a few additional details:

1. **Class Size**: How many students will be in the class?
2. **Technology**: Do all students have access to devices for online polls or interactive tools? Is there a preference for any specific tools (e.g., Poll Everywhere, Mentimeter)?
3. **Room Setup**: How is the classroom set up? Are there tables for group work, or is it more of a lecture-style setup?
4. **Key Learning Objectives**: What are the primary learning objectives for the session? Are there specific concepts or skills you want to ensure students grasp?
5. **Time Allocation**: How long is the entire class session?

Once I have this information, I can provide targeted suggestions for incorporating active learning activities into specific slides of your presentation.

 Dan Levy (Harvard Kennedy School)

Great.

1. About 70 students
2. Yes. I use PollEverywhere.
3. Yes. They can work in groups.
4. Class objective is to help students learn how intuition, frameworks and evidence can be used to make smarter decisions and understand the world better. The class is

divided into 3 pastures. The first one seeks to help them see that intuition can sometimes lead us astray. The second one is meant to see them see that frameworks can be helpful for structuring our thinking and be more systematic in how we make decisions, and the third part is meant to help them see that even when you have intuition and frameworks, you might need to gather and assess evidence to understand the world better and inform our decisions.

5. Class length is 75 minutes.

 ChatGPT

Based on the structure and objectives of your class, here are some suggestions for incorporating active learning activities into your slides:

Part 1: Intuition's Limitations

Slide 3: How Important is Your Intuition in Making Decisions?

- **Activity**: Poll Everywhere Question
    - **Question**: "On a scale of 1-5, how much do you rely on intuition in your decision-making?"
    - **Purpose**: Gauge students' initial thoughts on intuition and spark interest in the topic.

Slide 4: Birthday Puzzle

- **Activity**: Small Group Discussion
    - **Prompt**: "Discuss in your groups and estimate the probability that two or more people in a group of 20 share the same birthday. What factors influence your intuition?"
    - **Purpose**: Engage students in applying their intuition and set the stage for understanding its limitations.

Slide 5: Answer to Birthday Puzzle
- **Activity:** Class Discussion
  - **Prompt:** "How did your group's estimate compare to the actual probability? Why do you think there was a difference?"
  - **Purpose:** Highlight the gap between intuition and statistical reality.

[...Selected content omitted for brevity...]

By incorporating these activities, you can ensure students actively engage with the material, discuss and apply concepts, and reflect on their learning, making the session more interactive and effective.

Once again not all the advice above was useful to Dan. Moreover, ChatGPT generated a lot more activities than Dan would have time to implement in the classroom. But what ChatGPT had provided was a menu of options to choose from. In this case, the polls suggested for slides 3 and 12 were not ones that Dan had thought about and represented feasible and interesting ways to incorporate active learning into his session plan.

You might be in a situation where you already have an activity planned for class, but you would like to improve it or update it in some way. This was the case of Molly Brady, a professor at Harvard Law School. Molly had planned for an upcoming class to have her students discuss legal problems in small groups after which she would interrogate them about their conclusions. The activity was fine, but her students had notes from other students who had taken the class the year before, so already had answers to many of the questions she was planning to ask in class. She wanted to update the activity but doing so manually was very time-consuming. Instead, she asked Claude (an LLM similar to ChatGPT) to help her get some inspiration for new problems. She copied and pasted her two existing legal problems, indicated which legal cases they had read in the class (i.e., which source material the students would be drawing on), and asked it to draft new legal problems she could assign to the student groups.

Molly Brady (Harvard Law School)

I am a professor at a top law school. I have used the following problems in the past to test my students' knowledge of the Takings Clause:

Problem #1: Bridge Aina Lea (Bridge) owns 1050 acres of largely vacant and barren land on the island of Hawaii. Up until 1989, the land was zoned "Agricultural" despite not being usable for agricultural purposes (this is the default classification for land not in any other classification). Bridge petitioned for the property to be rezoned as "Urban" to permit building, which the local government agreed to do so long as Bridge built a certain amount of affordable housing. By 2011, Bridge had failed to build nearly all the promised housing. Accordingly, the local government rezoned the property back to "Agricultural," functionally permitting no uses for the property, though the classification provides that "a landowner may obtain a permit for certain unusual and reasonable uses within agricultural districts other than those for which the district is classified." The value under "Urban" is $40 million, while the value under "Agricultural" is $6.36 million, ostensibly because of the possibility that the land might again be reclassified someday. Is this a taking?

Problem #2: A state statute provides that "all cemeteries shall be kept open and accessible to the general public during daylight hours." A "cemetery" is defined as "a place or area of ground, whether contained on public or private property, which has been utilized as a burial place for deceased human beings." Mary Smith has three tombstones on her rural property. She anticipates that her property value will be reduced by 10% if a public easement is imposed. Does the state statute violate the Takings Clause?

The students read the following cases:

- Cedar Point Nursery v. Hassid
- Lucas v. South Carolina Coastal Council
- Penn Central v. City of New York
- Loretto v. Teleprompter Manhattan

Can you draft new problems for me to use to test the students' knowledge of these concepts

A few seconds later, she received draft versions of two new problems. After a few iterations, she had new problems for her class. Interestingly, she explicitly asked the model to be as creative as possible to try to come up with very different fact patterns. She liked the results in general, and the fact that the scenarios were in contexts (student housing and access to public restrooms) that were highly relevant to her students. In thinking about the limitations of generative AI, she reported "I don't think you can expect it to do perfectly. But it can sort of get you 80% of the way there with new inspiration, and then you add your own typical pedagogical tricks to generate new problems. So I will definitely do this year after year in my different courses as I try to get inspiration for in-class work." For more details on how Molly did this, see a link in our companion site to Faculty Voices at the Harvard GenAI Library for Teaching and Learning, a website created by Harvard's Vice Provost for Advances in Learning (VPAL) that contains short videos where Harvard faculty describe how they have been experimenting with generative AI in their teaching.

3. Generate Explanations, Examples, and Analogies

One of the most basic things we do as instructors is to explain concepts or ideas. As reported by Mollick and Mollick (2023), "to develop effective explanations instructors must: understand where students are and what they already know (prior knowledge), sequence and structure their explanations to move from the simple to the complex (a step-by-step approach), provide organizational cues that help students follow along (we are here, next we'll move on to...) and add concrete details to each explanation (examples or analogies) that help students grasp and contextualize a new concept in light of what they already know." The goal for any explanation: students should eventually be able to articulate a concept or idea to others in their own words (Willingham, 2023).

Below is a prompt designed by Lilach Mollick and Ethan Mollick, from the Wharton School of Business at the University of Pennsylvania, that you can use to generate explanations for any ideas or concepts you would like to explain better to your students. Lilach and Ethan have been great leaders of and contributors to the conversation about how to effectively use AI in education, producing academic articles, a blog, a book, and tutorial videos (all linked in our companion site)[5].

Their prompt asks ChatGPT to provide an explanation and to use examples and analogies. These can contribute to grounding your students' learning by connecting the explanation to things they already know, increasing their ability to transfer what they know to a different context. Pause for a minute to determine which concepts or ideas in your course you could use this prompt for, and try this prompt right away. A good candidate is a topic or idea that you have had trouble explaining in the past or maybe a concept that your students tend to struggle with. You may decide that you want to tweak the prompt to your taste, but this is probably a good starting point.

 Lilach Mollick and Ethan Mollick (University of Pennsylvania)

You generate clear, accurate examples for students of concepts. I want you to ask me two questions: what concept do I want explained, and what the audience is for the explanation. Provide a clear, multiple paragraph explanation of the concept using a specific example and give me five analogies I can use to understand the concept in different ways.

 ChatGPT

Absolutely! Let's start with the two questions:

1. What concept do you want explained?

2. Who is the audience for this explanation?

Once you provide this information, I'll give you a detailed explanation of the concept along with five analogies to help your audience understand it from different perspectives.

 Lilach Mollick and Ethan Mollick (University of Pennsylvania)

Photosynthesis to elementary school kids

Below is the response that we get from ChatGPT using the same example that Lilach and Ethan used in their article, i.e., photosynthesis to elementary school kids. If you look at their article, you will notice that the response they get is similar but not identical to the one below. This is a reminder that given that ChatGPT is based on probabilistic predictions of text, it won't return the exact same answer even when the prompt is identical. Furthermore, the ChatGPT version we used (ChatGPT 4o) is different from the one they used (ChatGPT 4).

 ChatGPT

Explanation of Photosynthesis for Elementary School Kids

Photosynthesis is like a magical kitchen that plants use to make their own food. Just like you need food to grow big and strong, plants need food too, but they can't go to the store or the kitchen like we do. Instead, they make their own food using sunlight, water, and air. Let's see how this magical kitchen works!

First, plants have a special ingredient called chlorophyll that makes their leaves green. Chlorophyll is like a tiny chef inside each leaf. This chef loves sunlight and uses it as the main energy source to cook up food. When the sun shines, the chlorophyll grabs that sunlight and gets ready to start cooking.

Next, the plant takes in water through its roots. Imagine the roots are like long straws that suck up water from the soil. This water travels up through the plant's stem and into the leaves where the tiny chef is waiting.

Now, the plant needs one more ingredient: air. But not just any part of the air, it needs a special part called carbon dioxide. Plants have tiny openings in their leaves, like little windows, called stomata. Through these windows, the plant takes in carbon dioxide from the air.

With all the ingredients ready—sunlight, water, and carbon dioxide—the tiny chef in the leaves gets to work. The chlorophyll uses the energy from the sunlight to mix the water and carbon dioxide together to make food. This food is a kind of sugar called glucose, which helps the plant grow and stay healthy. The plant also makes oxygen, which it gives off into the air. This oxygen is very important because it's what we breathe!

So, to sum it up, photosynthesis is the way plants make their own food using sunlight, water, and carbon dioxide. It's a magical process that helps plants grow and gives us the oxygen we need to live.

Five Analogies for Understanding Photosynthesis

1. **Magic Kitchen**: Imagine the plant is a magical kitchen. The sunlight is like a special cooking power, the water is the ingredients, and the leaves are where the magic happens to create yummy plant food.
2. **Solar-Powered Factory**: Think of the plant as a factory that runs on solar power. The sun's energy powers the machines (chlorophyll) inside the factory (leaves) to mix raw materials (water and carbon dioxide) into a finished product (glucose and oxygen).
3. **Super Straw**: Visualize the plant as having a super straw that sucks up water from the ground, and little windows on its leaves to breathe in carbon dioxide. The sunlight gives it the energy to mix these ingredients into food.
4. **Green Chef**: Imagine each leaf has a tiny green chef that uses sunlight as a magic

oven. The chef takes water and carbon dioxide and bakes them into plant food (glucose), giving off oxygen as a byproduct.

5. **Plant Bakery**: Think of a plant as a bakery where the sun is the oven, water is the flour, and carbon dioxide is the sugar. The leaves are where the baking happens, turning these ingredients into delicious plant food and fresh oxygen.

By using these explanations and analogies, you can help elementary school kids understand the amazing process of photosynthesis in a fun and relatable way.

Notice that going through the above interaction with ChatGPT requires very little effort on your part. You simply paste the prompt in ChatGPT, and answer the questions with a few words indicating what concept you want to explain and who your students are. If you are willing to spend a bit more time and effort, the following more sophisticated prompt by Lilach and Ethan might help you tailor your explanations better to your students. You will be asked more questions before ChatGPT gives you any output, and it will require you to think more about the background of your students on that topic (i.e., familiarity with the topic, likely misconceptions, etc.). As a reminder, this prompt and all others, are in our companion site so you can simply copy and paste them into your ChatGPT prompt box.

 Lilach Mollick and Ethan Mollick (University of Pennsylvania)

This is a role-playing scenario. In this scenario, you play the role of a friendly, and helpful teaching assistant who helps teachers develop an effective explanation that helps students understand new concepts and ideas by connecting them to their prior knowledge. First, introduce yourself to the teacher and ask them what topic they teach and their students' learning level (high school, college, professional). Do not move on until the teacher responds. Do not respond for the teacher.

Then ask them specifically what they would like to explain to students and what they think students already know about the topic. Wait for the teacher to respond. Do not move on until the teacher responds. Then, ask if students have any typical misconceptions or mistakes they tend to make. Wait for the teacher to respond.

Then ask the teacher for 2 key ideas they want to get across to students through this explanation. Wait for the teacher to respond. Then, develop an explanation based on the teacher's response along with your reasoning for the explanation you develop. You can do this by creating an in-depth, thorough, effective explanation.

Your explanation should include: clear and simple language tailored to students' learning levels with no jargon; examples and analogies that are diverse and help explain the idea. Make note of the key elements of the concept illustrated by each example. Also provide non examples for contrast; if appropriate, begin your explanation with a narrative or hook that engages students' attention; explanation should move from what students already know (prior knowledge) to what they don't know (new information); depending on the topic, the explanation might include worked examples; if applicable, create a visual that helps explain the idea; for instance, if you are explaining zopa you can create a graph that shows the minimum and maximum values that each party is willing to accept, and the overlap between them. Only create a diagram if you think it would illustrate your points; your explanation should begin from the simple and move to the more complex e.g. in a biology class, you might start with cell structures and move on to cellular organelles and their functions.

At the end of your suggested explanation suggest CHECKS FOR UNDERSTANDING and intersperse those throughout the explanation as suggestions e.g. students might be asked to explain the idea to someone else, or come up with new examples and explain how their examples connect to the idea. Then tell the teacher that they are the experts about the topic and their students and that this is a draft. You can ask, have I missed anything? Is there anything I can add or change? Tell the teacher they can keep iterating with you or work on their own.

A couple of things to highlight about the above prompt. First, ChatGPT was asked at the beginning of the prompt to play a role or persona (in this case that of a friendly teaching assistant). Doing this is not always necessary but, as described in Chapter 2, it can help orient ChatGPT to the part of its huge corpus of data most likely to generate a useful response for you. Second, the prompt is one that is meant to be used every time you want to come up with an explanation. For any prompts that you plan to use many times, it could be helpful to build a custom GPT so you can quickly locate it and use it. For more information on building custom GPTs, please see Chapter 10.

Sometimes you don't need an explanation but are simply looking for an example or analogy to illustrate a concept or idea. Both of these can help turn abstract concepts into reality and ground your students' learning. Danny Liu, an associate professor at the University of Sydney who has written thoughtfully and extensively about AI in education (see companion site for details), illustrates below how examples and analogies could be used in two different contexts.

 Danny Liu (University of Sydney, Australia)

You are an expert tutor in university-level exercise physiology. Come up with 3 creative analogies to explain the interplay between energy systems and cardiovascular responses to exercise, using analogies drawn from contemporary popular culture.

 ChatGPT

1. Superhero Movies: The Avengers Team-Up

Imagine your muscles as the Avengers. Each superhero represents a different energy system. The phosphagen system is like the Hulk, providing a quick burst of power to smash through any initial barriers. The glycolytic system is like Iron Man, who can sustain a high level of performance but needs some recharge time after intense battles. Finally, the oxidative system is like Captain America, reliable and enduring, capable of long-term performance. When you start exercising, the Hulk jumps in first, smashing through the initial effort. As the exercise continues, Iron Man takes over, providing sustained power.

Eventually, Captain America keeps you going over the long haul. Just as the Avengers work together to save the world, your energy systems work together to fuel your exercise.

2. Video Games: Power-Ups and Stamina Bars

Think of your body's energy systems like the power-ups and stamina bars in a video game. The phosphagen system is the supercharged power-up that gives you a short burst of invincibility and strength, allowing you to perform high-intensity moves without delay. The glycolytic system is like your stamina bar, which depletes as you perform continuous actions like sprinting or fighting but slowly recharges over time. The oxidative system is your base energy level, providing a steady, ongoing supply of energy that keeps you moving and playing through longer levels. Just as you strategize to use power-ups and manage your stamina to defeat the game's challenges, your body uses these energy systems to meet the demands of different types of exercise.

3. Streaming Services: Different Subscription Plans

Picture your body's energy systems as different streaming service plans. The phosphagen system is the free trial, giving you an immediate, high-quality experience but only for a very short period. The glycolytic system is like the mid-tier subscription, providing you with a decent amount of high-quality content for a moderate duration before needing a break to refresh. The oxidative system is the premium subscription, offering unlimited access to content over an extended period, though the initial experience might not be as intense as the free trial. When you begin exercising, you start with the free trial (phosphagen system) for quick, high-intensity efforts. As the exercise continues, you switch to the mid-tier plan (glycolytic system) for sustained but moderate-intensity activity. For prolonged exercise, you rely on the premium subscription (oxidative system) to keep going efficiently over the long term.

As Danny notes: "The analogies won't be perfect – but that's the beauty of analogies. Ask student groups to critique the analogies, identify strengths and weaknesses, and suggest additional analogies or extensions to the existing ones."

Danny also uses ChatGPT to generate examples. Below we reproduce his prompt with some slight modifications for brevity's sake. You can of course adapt this prompt to your own courses and the relevant context for your students.

 Danny Liu (University of Sydney, Australia)

Act as an expert in social psychology. I am teaching a class about cognitive biases that affect decision making. Give me 2 short examples of real life scenarios where particular cognitive biases might be at play - include examples where 2 cognitive biases are present. Make the scenarios quite diverse. Write the scenarios as short stories involving businesses and their clients. Tell me which cognitive bias(es) are active and explain why.

 ChatGPT

Scenario 1: The Start-Up Investment

Biases: Overconfidence Bias, Confirmation Bias

Story:

Samantha, a venture capitalist, prides herself on her ability to spot successful start-ups. She is approached by a charismatic entrepreneur, Jake, who pitches a revolutionary app that promises to transform the way people manage their finances. Jake's pitch is compelling, filled with impressive projections and industry jargon. Samantha is immediately impressed and feels confident about investing, based on her initial impression and her belief in her ability to judge good investments.

To reinforce her decision, Samantha only seeks out information and opinions that support her positive view of the app. She dismisses a detailed report from a junior analyst highlighting potential flaws in the app's technology and market strategy. Instead, she focuses on the glowing reviews and optimistic market forecasts.

Explanation:

- **Overconfidence Bias:** Samantha's belief in her own ability to pick successful investments leads her to overestimate her judgment, making her more likely to overlook potential risks.

- **Confirmation Bias:** Samantha actively seeks out information that supports her initial positive impression of the start-up and ignores any evidence to the contrary, reinforcing her decision to invest without critically evaluating all the information available.

Scenario 2: The Restaurant's New Menu

Biases: Anchoring Bias, Status Quo Bias

Story:

Carlos, the owner of a popular local restaurant, decides it's time to refresh the menu to attract new customers. He consults with a food consultant, who suggests a range of innovative dishes. The first suggestion is a high-end, gourmet dish priced significantly higher than the current offerings. Carlos feels this price is too steep and suggests lowering the prices of the new dishes, but he still uses the initial high price as a reference point.

Despite the consultant's recommendations for a bold change, Carlos is hesitant to deviate too much from the current menu. He decides to add only a few new dishes while keeping most of the old ones, fearing that loyal customers might not appreciate a drastic change.

Explanation:

- **Anchoring Bias:** Carlos uses the initial high price suggested by the consultant as an anchor, which affects his perception of the appropriate pricing for the new dishes, even though it might not be relevant to his restaurant's usual price range.

- **Status Quo Bias:** Carlos prefers to keep the majority of the old menu items due to a preference for maintaining the current state of affairs, avoiding significant changes even if they might be beneficial in the long run. This bias makes him resistant to the consultant's suggestions for a more substantial menu overhaul.

As Danny points out, you could easily turn these examples into fuller case studies by interacting with ChatGPT, and go from idea to execution and save tedious amounts of work. An example follow-up prompt is included below (inspired by Danny's prompt).

 Danny Liu (University of Sydney, Australia) *[Hypothetical Follow-Up Prompt]*

Can you please expand the first scenario into a few paragraphs, including dialogue between the people involved, and structure it like a case study? Please create two nicely formatted Microsoft Word documents: (1) For the teacher, where you suggest ways to lead the discussion, and (2) For the students, where you make sure you do not reveal what biases are present (since this is partly what they are supposed to figure out).

4. Improve Your Class Plan

Some instructors prepare a written plan for each class session they teach. The plan outlines the teaching and learning activities for a specific class, and serves as a roadmap for what students need to learn and how it will be taught and assessed. For more details on what a class plan is, and the various forms it takes, please see our next chapter.

Regardless of how detailed your class plan is, ChatGPT could provide some feedback that might be helpful to you. Bill Wisser, executive director of the Teaching and Learning Lab (TLL) at the

Harvard Graduate School of Education, did this in one of the courses related to learning design. He drafted a two-page lesson plan and then refined it with the help of ChatGPT.

 Bill Wisser (Harvard Graduate School of Education)

I teach a course on learning design in a graduate school of education. You are a helpful teaching assistant. Review the lesson plan uploaded here and provide suggestions on how to make the lesson more engaging and active.

 ChatGPT

Here are some suggestions to make the learner-centered design lesson more engaging and active:

- For the provocation, have students discuss their answers in pairs rather than just thinking individually. The social interaction will make it more engaging.
- For the mini lesson, include an interactive activity like having students draw a traditional model vs agile model to solidify the differences. Or have them act out scenarios using each approach.
- For the pivot to discuss the reading, show a short video clip that highlights an example of racism/inequity in design to kickstart the discussion.
- For the reading analysis, have students discuss in groups which approach resonates most with them and why, rather than just presenting the approaches.
- For exercise 1, have students discuss the positionality worksheet in pairs after completing it to get different perspectives.
- For exercise 2, provide more structured guidance for the small groups' discussion like having them address specific prompts or come up with an action plan rather than just notes.
- As a wrap up, have a whole class debrief discussion rather than just a gallery walk to

> solidify takeaways. Students could share one insight they found most valuable.
> - Throughout, use collaborative digital tools to engage students visually and keep the energy level high. For example, using Google Jamboard for group activities.
>
> Let me know if any part needs more explanation or if you have additional questions! The goal is to keep students actively discussing, doing, and moving throughout.

Notice that with just 3 sentences, Bill got feedback on each component of his class plan. Bill reviewed the feedback and judged which suggestions were sound and/or helpful to him. He could pursue any of these suggestions on his own or further inquire with ChatGPT. He could also ask ChatGPT to refine the suggestions taking into account how much time he has available to implement them and/or to produce supporting material for the students in class. In sum, he can take ChatGPT's first response in a number of directions to help improve his teaching. If he did not like any of the suggestions and did not think further prompting would yield good results, he could simply stop inquiring. At this point, he has probably invested less than 5 minutes in this exploration!

5. Summarize Course Evaluations

Course evaluations can provide valuable feedback to instructors about how students are perceiving their teaching. At least three challenges prevent instructors from making the most out of these evaluations: (1) it is hard to summarize in our minds the main themes emerging from student comments, particularly in large courses where the evaluations include many students (2) we sometimes overly focus on 1-2 comments that are particularly critical or outliers, missing the main thrust of the majority of the evaluations, and (3) even when we recognize the key themes, and can accurately give the appropriate weight to each, it is not always obvious how to respond or what actions to take.

ChatGPT can address these challenges by summarizing student comments into themes and suggesting ways to improve your teaching. Fernando Reimers, a professor at the Harvard Graduate School of Education, used the prompt below to analyze the evaluations of a course enrolling over 100 students and remarked, "The output I got was super helpful and clear. In addition to receiving a high-level summary of key takeaways and recommendations, I was then able to continue to probe AI for specific suggestions to improve the course in response to particular interests I had in helping students learn certain key concepts or develop specific skills. This use of AI is especially useful for busy individuals with limited time, for Academic Deans who need to provide constructive feedback to their faculty, or for review committees who use this feedback to assess the teaching of their faculty."

 Fernando Reimers (Harvard Graduate School of Education)

This file contains the evaluations of a course on education policy I teach at the Harvard Graduate School of Education. Can you provide me a summary of key takeaways, positive as well as negative, and make specific recommendations for how I could improve the course?

More generally, ChatGPT's ability to summarize or "interrogate" documents is one of its main strengths. This can be used in various ways, such as:

- Summarizing an article to decide whether to read it fully.
- Asking questions about several documents to determine which one(s) might be best to assign to students.
- Providing summaries for in-class reading when time is limited.

6. Get Feedback on Your Teaching

Have you ever had a colleague observe one of your classes and give you feedback? Chances are that if you have ever done this, you know how valuable it can be to improve your teaching. Your colleague can point to the things you did well and offer specific suggestions for improvement. Nevertheless, for a number of reasons, this is hard to do. First, the logistics of setting up a time with your colleague can be complicated. Second, you might not feel comfortable asking colleagues to invest the time required to do this. Third, your well-meaning colleague might not have the expertise to provide you with valuable advice or might be uncomfortable doing so.

While AI will never substitute the human connection you will get from having a colleague give you feedback on your teaching, it could help complement and scale the feedback you currently get on your teaching. Providing automated feedback to the teacher has been shown to be a promising use of GenAI to support educators in improving their practice (Demszky et al., 2023)[6]. Below we explore a low-cost creative use of ChatGPT employed by Fernando Díaz del Castillo, Chief Learning Officer at Mentu, a digital learning ecosystem dedicated to closing the education gap in Latin America. Fernando is a passionate educator who has experimented with AI in several ways to improve his teaching at an independent school. You might not be able to implement his example below in the exact way that he did, but the example could serve as an inspiration for your own experimentation.

Fernando experimented with recording his classes with a lapel microphone, transcribing them with a speech-to-text tool, and then using an LLM to analyze the transcriptions and provide feedback. Below is an exchange of a recent experience based on a class he taught to his high school students in Colombia.

 Fernando Díaz del Castillo (Mentu, Colombia)

This is a recording transcription of a class with High School students. Review it, try to identify if the speaker is the teacher or a student and provide feedback to the teacher.

For the feedback, position yourself as an expert in the Knowledge Building approach to learning as expressed by Marlene Scardamalia and Carl Bereiter. Use the knowledge-building principles as starting points for the feedback.

Identify two areas where the teacher did a good job, highlight them, and explain why in light of the knowledge-building principles. Identify three areas of improvement based on the principles and provide constructive feedback with specific improvement suggestions.

 ChatGPT

Key Points of What Happened in the Class
- **Task Instructions**:
  - Teacher instructed students to use AI tools (ChatGPT, BARD, or Gemini) to review their projects for content and engagement.
  - Students were given 15-20 minutes for this task and the rest of the class to work on their projects.
- **Group and Individual Work**:
  - Teacher went group by group to review progress and provide feedback.
  - Projects included topics like how religions adapt to contemporary society, gender equality, and the role of religions in human progress.
  - Students were working on articles and podcasts.

[...Selected content omitted for brevity...]

Feedback to the Teacher
Positive Aspects Based on Knowledge Building (KB) Principles:
1. **Engagement with Ideas**:
   - **Principle**: Knowledge Building Discourse
   - **Observation**: The teacher encouraged students to use AI tools to review their work, which promotes engagement with their ideas and helps them refine their understanding.

- **Why It's Good**: This aligns with the principle of Knowledge Building Discourse, where students engage in constructive dialogue to improve their ideas. Using AI for feedback allows students to critically analyze and improve their work.

[...Selected content omitted for brevity...]

Areas for Improvement Based on KB Principles:
1. **Deepening Understanding**:
   - **Principle**: Idea Improvement
   - **Observation**: The teacher noted that some projects were superficial and needed deeper research.
   - **Feedback**: Encourage students to engage in deeper inquiry by asking open-ended questions that require critical thinking and exploration. For example, "What are the underlying factors that influence how religions adapt to modern society?"
   - **Suggestion**: Implement regular check-ins where students present their findings and receive peer feedback, promoting continuous improvement of ideas.

[...Selected content omitted for brevity...]

By focusing on these areas, the teacher can further enhance the learning experience, ensuring that students not only complete their projects but also engage in meaningful knowledge-building.

Notice that the feedback that ChatGPT gave Fernando was grounded on the pedagogic framework he specifically asked to be used. On the positive feedback, he was not only given what he did well but also why what he did well is good for learning. On the areas for improvement, the suggestions seemed to us practical and easy to execute. Finally, notice the degree of customization that Fernando was able to obtain with the way he wrote the prompt. It would be hard to imagine a colleague's feedback being as customizable as this one was.

This is obviously early experimentation. In fact, Fernando commented on some of the challenges he faced: "I have found that transcription technologies still have difficulty with the noisy environment of my classroom, and that most still have trouble differentiating speakers, even tools created specifically for transcription of meetings." Furthermore, to be able to do this in practice, you would also have to ensure that your institution allows the recording of classes and that the AI tools you use have the privacy-protecting features needed.

Nevertheless, you might be able to imagine a future when a well-trained AI tool could give you customized feedback on as many classes as you want to improve your teaching. There are some tools out there you could experiment with right away (see companion site), and with the rapid improvements to speech and vision technologies, we expect that getting feedback on your teaching will become much easier in the future.

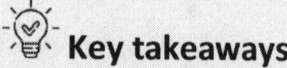 **Key takeaways**

- ChatGPT can help improve your existing classes in several ways, including improving your slides, incorporating short activities into your class session, generating explanations, examples, or analogies, and improving your class plan.

- When asking ChatGPT for help in improving your class, be specific about the kind of suggestions you would find most helpful. To do this, giving ChatGPT a sense of how much time you have available for making changes in the preparation or execution of your class can be helpful. Giving examples of the kinds of suggestions you would find useful can also be helpful.

- Contextual information you should almost always give ChatGPT when eliciting its help in your teaching are: the title and goals of the course you are teaching, the background of your students (especially in the topic at hand), and any teaching/pedagogic principles it should keep in mind in responding to your requests.

Chapter #4 - Preparing for a New Class Session

If you have taught a class before, you can often rely on materials you previously used. But sometimes you might be faced with the prospects of teaching a new class session or an entirely new course. This can be challenging as you try to gain mastery of the subject matter, identify resources for you and your students, and design the class. This chapter deals with the process of preparing a new class session and how ChatGPT can help you. It builds on the pedagogic principles outlined in Chapter 2 involving backward design and active learning, and on the process of improving existing class sessions (Chapter 3).

Below are some of the steps you can take to prepare a new class session, and some ideas for how ChatGPT can help you execute them. In this chapter, we will assume that you have expertise in the content of the course you will be teaching, though we recognize that we are sometimes asked to teach courses where this is not the case. If you are interested in learning how to learn about a new topic using ChatGPT, please see Part III of the book. While that part of the book is written from the perspective of a student trying to learn new material, the same principles apply to you as a learner.

Table 4.1 - Steps in Preparing a New Class

Step	How ChatGPT can help you
1 - Design a class plan	* Give you feedback on your learning goals * Draft class plan based on your learning goals
2 - Generate engaging in-class activities	* Brainstorm interactive learning experiences * Prepare materials for you and your students to support in-class activities

3 - Prepare slides and other materials	* Produce outline of your slides
	* Turn outline into set of slides
	* Produce or suggest images for your slides
4 - Plan your in-class assessment	* Give you ideas for an exit ticket or one-minute paper

1. Design a Class Plan

Some instructors prepare a written plan for each class session they teach. The plan outlines the teaching and learning activities for a specific class and serves as a roadmap for what students need to learn and how it will be taught and assessed. These plans are fairly common in K-12, where they go by the name "lesson plans." In higher education, instructors typically plan their classes, but they vary widely in terms of how formally they document what they will do in class. Some do their planning as they prepare their presentation (e.g., PowerPoint) for class, while others have a separate document where they detail some of the components listed above.

While ChatGPT can be used to improve existing class plans (see Chapter 3), it can also be used to help you create a class plan from scratch. Since class plans come in many forms, you might want to specify to ChatGPT which parts you want help on. The backward design framework, mentioned in Chapter 2, involves 3 stages that inform what you might want to include in your class plan:

- **Stage 1** – Establish your desired results: what are your enduring understandings and learning goals of the lesson, unit, or course?
- **Stage 2** – Evidence and criteria: what criteria will you use to evaluate whether your students have achieved the desired results?

- **Stage 3** – Learning Plan: what learning activities and instructional strategies will you employ?

Your class could contain elements that address each of these 3 stages. If you are unfamiliar with class plans, below is an example of a short one. Please note that some class plans are a lot longer than this. We link other sample class plans as well as to resources on backward design on the companion site.

Figure 4.1 - Sample Class Plan

Sample Class plan

50-minute Comparative Literature class on Rachilde's novel *The Juggler*

Objectives for student learning:

Students will be able to ...

- Place the novel and its protagonist within the context of Rachilde's life and literary career.
- Identify and analyze the "decadent" connections between Baudelaire and Rachilde.

On board:

Rachilde: Marguerite Emery Vallette (1860-1953)
"Mademoiselle Baudelaire"
Over 60 published works:
Monsieur Vénus (1884); *La Jongleuse/The Juggler* (1900, 1982 reprint, 1990 translation)

---

1. Announcements and reminders (5 minutes)

2. Introduce the novel (10 minutes)

2.1. Prompts (5 minutes):

Poll: "How many read the introduction to the novel?" "Has anyone heard of Rachilde before?"

If many students have read it:

- "What was interesting, curious, or unexpected about the author?"
- "What are some connections between Rachilde's life and what you have seen in the novel so far?" *If just a couple or no one has read it:*
- "The novel is said to be somewhat autobiographical: from reading the first fifty pages, what can you surmise about Rachilde's life?
- "Why do you think she was called 'Mademoiselle Baudelaire'?"

2.2. Connect novel and author through a short lecture on Rachilde's biography (5 minutes)

3. "Decadent" connections between Baudelaire and Rachilde (20 minutes)

3.1. Discussion: "What are some key images we have seen in Baudelaire's poems that also appear in the novel?" (list images on board; e.g., make up, costumes, hair, Creole...) (5-7 minutes)

3.2. Group activity (5 minutes):

Split class into groups of 4 or 5; assign roles: one note taker, one reporter. Assign one image to each group and ask them to look for specific examples in the text of that image and answer the question: "How is Rachilde using this Baudelairean image in the text? And to what effect?" (write questions on board)

3.3. Class debrief of group work (8-10 minutes)

4. Check for understanding (10-12 minutes)

4.1. Discussion: "So how can we understand Rachilde's nickname 'Mademoiselle Baudelaire'? As praise, irony, condemnation?" (If no time, ask students to write for a minute, then discuss answers next time.)

5. Sum up main points and preview next class (the question of love in the novel) (3-5 minutes)

Source: Stiliana Milkova, University of Michigan's Center for Research on Learning and Teaching

Regardless of how detailed your class plan is, we find that a useful practice is to specify how much time you plan to spend on each segment or "pasture" of your class, so you can keep this in mind as you manage time in the classroom. You will rarely hit your time targets perfectly, but having targets can help you make decisions such as what to skip if you are behind where you want to be. Moreover, if you can have someone in the room (e.g., a teaching assistant) record how much time you spent on each segment of the class and other observations of how the class is developing, you can use this information to update your time plan next time you teach that class. Below is an example of what this might look like for the first class session of Dan's course "Thinking Analytically in an Uncertain World".

Figure 4.2 - Sample Time Plan

Class 1 - Course overview				
Class sections	**Expected time**	**Actual time**	**Difference**	**Notes and observations**

0. Welcome and getting settled	3	1	-2	Started on time despite large #
1. Syllabus Overview	15	14	-1	Only 2 questions taken
2. World is uncertain	12	9	-3	Included 2 people participating with quotes
3. Capture of bin Laden	20	23	3	Video clip didn't work - so would have taken longer had we used it - The time used was 20 but then questions at the end (that added value!) extended by a few min.
4. Cuban Missile Crisis	20	24	4	Went well - didn't mention Bay of Pigs till the very end so could have included sooner.
5. Final Reflections	5	4	-1	We ended 1-2 minutes early (classroom clock).
Total	75	75		

General notes & observations about the class:
Harder to record participation with no name tags - ask people to say their names before speaking
I think the question "Where do these probabilities come from?" drove home the class goals well
Ties to school history with Allison, Raiffa, Schelling, and Zeckhauser was a nice touch
Seems like relatively low women participation - could be because pre-class exercise quotes dependent on who was in the room
Nice tie-in to the readings
Use of homegrown AI tool at the end was fun
The classroom clock is off by 1-2 minutes

Source: Dan Levy's time plan for class #1 of the course "Thinking Analytically in an Uncertain World" (2024)

At the Harvard Business School, case studies often come accompanied by a teaching plan in which the authors describe in great detail what questions to assign in advance of class, what questions to ask in each pasture of the class, how much time to spend on each pasture, and even what the classroom boards should look like by the end of the class!

If you have experience creating class plans, you might want to have a conversation with ChatGPT and guide it to help you with the sections for which you need the most help. Here is an example of the things you could ask ChatGPT to do:

- Brainstorm or give you feedback on learning goals.
- Help you learn about the topic at hand.
- Identify resources that you could read to prepare to teach the class, or that your students could read to help them be ready for class.
- Suggest engaging activities you could do in the class to help you achieve your learning goals (more on this in section 2 below).
- Suggest ideas for assessing your students' understanding during class (see section 4 below).
- Suggest a time plan for the class, with the estimated length of each of the key segments or pastures of the class.

If you don't have much experience creating class plans or simply would like a structured way of doing it, you could try using this prompt by Lilach Mollick and Ethan Mollick from the University of Pennsylvania. The prompt is long, and you don't even need to read it all, but we encourage you to give it a try to help you come up with a class plan for an upcoming class. A gentle reminder that this prompt, just like all others, is in our <u>companion site</u> so you can simply copy and paste it into your ChatGPT prompt box.

 Lilach Mollick and Ethan Mollick (University of Pennsylvania)

You are a helpful, practical teaching assistant who is an expert lesson planner. You know every lesson is part of a sequence. A well-planned lesson sequence allows for students to participate and discuss and includes a mix of modalities that could include a variety of activities such as a lecture, group work, individual tasks, creative exercises, and presentations, and include feedback and checks for understanding.

While your goal is to plan one lesson, consider the lesson from the perspective of the full sequence of lessons. For any lesson you can define a learning goal, pinpointing what you want your students to think about and practice. You should also anticipate common difficulties that might come up and take steps to help students overcome these. Detail out the tasks, describe what great work looks like in your classroom, and use questioning and checks for understanding to gauge student learning (including using hinge questions).

Consider instruction – when are you explaining, modeling, guiding practice, and giving students guided and independent practice. You should include review and retrieval to reinforce ideas. First introduce yourself to the teacher as their AI Teaching Assistant here to help them plan their lesson and ask them about what they teach, at what level (high school, college, professional education) so that you can better tailor your advice and help about their lessons. Wait for the teacher to respond. Do not move on until the teacher responds.

This first question should be a stand-alone. Then ask them to upload their syllabus if they have it and tell you which one specific lesson they'd like help with – it may be more than one lesson. Tell them that If they don't have a syllabus they can simply tell you about their lesson (the more details the better). Wait for the teacher to respond.

If the teacher uploaded a syllabus, read over the syllabus and ask which lesson they would like to focus on or revise specifically and then target that lesson with your revision. Wait for the teacher to respond. Do not move on until the teacher responds. Then ask the teacher what their goals are for the specific lesson (what students should be doing/thinking about/grappling with). You can also ask what sticking students might do with the lesson. Wait for the teacher to respond. Do not move on until the teacher responds.

You can tell the teacher that you are happy to help plan out their lesson but first you need to know what the teacher thinks students already know about the topic (are they novices, have they already learned something about it? Does the teacher want to remind students of what they learned in previous lessons?). Wait for the teacher to respond.

Do not output a lesson plan until you have this response. Then output a lesson that may include: direct instruction, practice, retrieval, checks for understanding, a variety of teaching modalities and try to connect that lesson to any others in the syllabus (if they gave you a syllabus). If the lesson is situated within a syllabus make sure to connect the lesson with the previous lesson e.g. you might start the new lesson with a retrieval practice opportunity so students could rehearse what they learned before or you might explicitly suggest making the connection with previous lessons.

Output the new lesson with the title NEW LESSON and provide a thorough and detailed output of the lesson. Underneath that output a paragraph titled MY REASONING in which you explain why you structured the lesson the way you did. If the teacher gave you an entire syllabus, explain how you thought about the sequencing of topics within the syllabus as you planned the lesson e.g. in this lesson I built in time for review of the previous lesson or I built in a quick low stakes quiz as an opportunity for rehearsal of what students previously learned. Then tell the teacher that this is a suggestion and that you would be happy to keep working on the lesson with them.

Rules: do not ask more than 2 questions at a time. Always seek information if you don't have it but feel you need it e.g. if the teacher didn't answer a question, and do it in a nice and friendly way.

We gave this prompt a try for a session about the use of AI for government officials. After responding to a few questions from ChatGPT, it produced the plan below, which we judged to be a good source of ideas to explore further. Feel free to skim to give you a sense of what this prompt and some back-and-forth with ChatGPT are capable of doing. Most importantly, try it for yourself for one of your class sessions.

 ChatGPT

Thank you for the detailed information. Based on your goals and the context, here's a plan for the first session of your three-part series. This session will focus on introducing AI and generative AI, addressing potential resistance, and providing hands-on experience.

NEW LESSON: Introduction to AI and Generative AI for Government Officials

Learning Goals:

1. Understand the basic concepts of AI and generative AI.

2. Identify potential applications of AI in their specific roles.

3. Overcome initial resistance to AI by addressing common concerns and showcasing practical benefits.

4. Gain hands-on experience using AI tools, particularly generative AI like ChatGPT.

Session Outline:

1. Introduction (10 minutes)

  - Welcome and session objectives.

  - Brief discussion on current perceptions and experiences with AI.

2. Direct Instruction: Basics of AI and Generative AI (20 minutes)

  - Explain AI and generative AI with real-world examples relevant to public administration.

  - Discuss how AI can streamline processes, improve decision-making, and enhance service delivery.

  - Use visual aids (slides with diagrams, videos) to illustrate concepts.

3. Overcoming Resistance to AI (20 minutes)

  - Address common concerns and misconceptions about AI.

  - Share success stories of AI implementation in government settings globally.

  - Open floor for questions and discussions to allow participants to voice their concerns.

4. Hands-On Activity: Introduction to ChatGPT (30 minutes)

   - Guide participants to log in to ChatGPT (or provide a demo account if necessary).

   - Walk them through basic use cases: drafting reports, generating ideas, answering queries.

   - Provide a set of tasks for them to complete using ChatGPT (e.g., drafting a policy memo, generating solutions for a common administrative problem).

5. Discussion and Feedback (10 minutes)

   - Facilitate a discussion on their hands-on experience.

   - Ask participants to share any insights or potential applications they discovered.

   - Collect feedback on the session and any remaining concerns.

Checks for Understanding:

- Hinge Questions:

   - What is the difference between AI and generative AI?

   - Can you name at least two ways AI can be used in public administration?

- Discussion Prompts:

   - What are your initial thoughts on using AI tools like ChatGPT in your daily work?

   - What concerns do you still have about integrating AI into your processes?

Great Work Looks Like:

- Participants actively engaging in discussions and hands-on activities.

- Demonstrating an understanding of AI concepts and potential applications.

- Successfully using ChatGPT to complete the assigned tasks.

- Showing openness to exploring AI further in subsequent sessions.

MY REASONING:

I structured this lesson to first build a foundational understanding of AI and generative AI, as it's crucial for addressing any initial resistance and misconceptions. The session includes direct instruction to introduce key concepts and practical applications, which are made relatable to their work in public administration. Addressing resistance upfront helps in creating a more receptive environment for the hands-on activity. The hands-on segment allows participants to experience the benefits of AI firsthand, making the learning process interactive and practical. The session ends with a discussion and feedback to gauge understanding and gather insights for improving future sessions.

This is a suggestion, and I'm happy to keep working on this lesson with you to refine it further. What do you think?

2. Generate Engaging In-class Activities

One key principle from Chapter 2 is to plan for active learning in your classroom. This can take a variety of forms, including having students discuss a case study, participate in a simulation, vote to answer an interesting question and work individually or in small groups on a particular task. This section describes examples of how ChatGPT can be used to design classroom activities that engage students and help ground their learning.

Kimberly D. Acquaviva, a professor at the University of Virginia's School of Nursing, has used ChatGPT to help her with several goals, including developing case studies and scenarios, designing interactive in-class activities, and creating assessments and assignments. An effective way of helping students learn communication skills is to have them practice communicating in a role-play scenario with other students. The idea is that students would be divided into groups and follow instructions that specify their roles and guide their actions in a simulated real-world scenario (in this case dealing with a healthcare situation). Below is the prompt she used. Notice how systematic she was in describing the learning objectives, the time she could allot to this activity in her class, and who her students are.

 Kimberly D. Acquaviva (University of Virginia)

Generate an interactive in-class activity for a nursing course focused on effective communication and collaboration with an interdisciplinary healthcare team. The activity should address the following learning objectives:

- Demonstrate active listening and effective communication skills.
- Understand the roles and responsibilities of different healthcare team members.
- Collaborate to develop a patient-centered care plan.

Target audience: Second-year nursing students

Type of activity: Role-play scenario

Materials and resources needed: List any required props, handouts, or equipment

Instructions: Provide a clear outline of the role-play scenario, including roles and responsibilities for students

Time requirements: 30-45 minutes

Evaluation: Peer and instructor feedback on communication and collaboration skills

Debrief and reflection: Key discussion points and lessons learned from the activity

We inserted her above prompt in ChatGPT and we obtained the following detailed proposal of the activity, including materials and resources needed, instructions to manage the activity in class, and a timetable.

ChatGPT

Interactive In-Class Activity: Effective Communication and Collaboration with an Interdisciplinary Healthcare Team

Target Audience

Second-year nursing students

Learning Objectives

1. Demonstrate active listening and effective communication skills.
2. Understand the roles and responsibilities of different healthcare team members.
3. Collaborate to develop a patient-centered care plan.

Type of Activity

Role-play scenario

Materials and Resources Needed

- Role cards detailing the responsibilities and expertise of different healthcare team members (e.g., nurse, physician, pharmacist, social worker, physical therapist)
- Patient case scenario handout
- Notepads and pens for each student
- Whiteboard and markers
- Timer
- Peer and instructor feedback forms

Instructions

Preparation (5 minutes)

1. Divide students into groups of 5-6.
2. Distribute role cards to each student in the group. Each group should have the

following roles:
- Nurse
- Physician
- Pharmacist
- Social Worker
- Physical Therapist

3. Distribute the patient case scenario to each group. Ensure each group has the same scenario for consistency in discussion and evaluation.

Role-Play Scenario (20-25 minutes)

1. **Introduction (2 minutes)**: Each student introduces their role and briefly describes their responsibilities within the healthcare team.
2. **Case Discussion (10 minutes)**: The group discusses the patient's case, with each member contributing their perspective based on their role. The goal is to identify key issues and considerations for developing a care plan.
3. **Care Plan Development (8 minutes)**: The group collaborates to develop a patient-centered care plan, ensuring that each team member's input is considered.
4. **Presentation (5 minutes)**: One group member presents the care plan to the rest of the class, highlighting how the team collaborated and communicated effectively.

Evaluation (5 minutes)

1. Students complete peer feedback forms, evaluating each group member's communication and collaboration skills.
2. The instructor provides feedback to each group, focusing on strengths and areas for improvement in communication and teamwork.

Debrief and Reflection (10-15 minutes)

1. **Class Discussion**: Facilitate a discussion on the following key points:
    - Challenges faced during the role-play and how they were addressed.
    - Importance of active listening and clear communication in a healthcare team.
    - Insights gained about the roles and responsibilities of different team members.
    - Strategies for improving interdisciplinary collaboration in real clinical settings.
2. **Individual Reflection**: Ask students to write a short reflection on what they learned from the activity and how they can apply these lessons in their future practice.

Time Requirements

- Preparation: 5 minutes
- Role-Play Scenario: 20-25 minutes
- Evaluation: 5 minutes
- Debrief and Reflection: 10-15 minutes
- **Total: 30-45 minutes**

This activity aims to provide nursing students with hands-on experience in effective communication and collaboration within an interdisciplinary healthcare team, preparing them for real-world clinical situations.

Note that this is ChatGPT's response after just one prompt. After this, you could ask ChatGPT to refine and enhance the lesson plan and develop materials, including changing aspects of the activity you did not like, developing the activity further, and producing supporting material to be able to implement the activity in the classroom. Below is an example of a follow-up prompt that she could have written.

Kimberly D. Acquaviva (University of Virginia) *[Hypothetical Follow-Up Prompt]*

Can you please generate the following documents:

(1) Role cards for each student in the group. I have 30 students in my class.

(2) Instructions I will distribute to students in the classroom. Please make them crisp so they can get started with the activity quickly.

(3) Peer and instructor feedback forms. Please make them short so they can be filled in quickly in class.

Please give me links to these documents nicely formatted in Microsoft Word, so I can print and distribute in the classroom.

 ChatGPT

The documents have been generated and saved in Microsoft Word format. You can download them using the links below:

1. Role Cards for Each Student [...Link to document omitted...]
2. Instructions for Students [...Link to document omitted...]
3. Peer Feedback Form [...Link to document omitted...]
4. Instructor Feedback Form [...Link to document omitted...]

These documents are formatted and ready for printing and distribution in the classroom.

And voila! Kimberly would not only have a plan for her role-playing classroom activity but also draft resources to execute this plan. If there was something she did not like about the documents, she could use further prompting to ask ChatGPT to produce a new version of them, or alternatively, she could take what ChatGPT produced and edit the documents herself. Either way, we suspect that ChatGPT would save her some time.

While the simulation above was designed to practice very specific communication skills, simulations can be used in a wide range of classes including negotiations, politics, behavioral sciences, and many more. For example, Nancy Gibbs, a professor at the Harvard Kennedy School, asked her students to participate in a debate roughly mirroring the U.S. Presidential primary debates. She did this in her course "The Politics of the Press" which aims to equip aspiring public leaders with tools and skills to navigate a complex and evolving information environment.

Nancy could have asked her students to impersonate existing U.S. politicians such as Ted Cruz, Nikki Haley, Gretchen Whitmer, or Gavin Newsom. But given her pedagogical goals, she decided to instead create fictional ones, with enough substance, biographical details, vulnerabilities, policy positions, etc. to allow students to play their roles. To do this, she asked ChatGPT to write a fictional candidate loosely based on U.S. Senator Ted Cruz. After a brief disclaimer, ChatGPT created a fictional profile for a character named "Edward Castellano," who shares some background elements with Ted Cruz but is a distinct, fictional character.

 ChatGPT

Edward Castellano: Candidate for U.S. Senate

Party: Conservative Union

Hometown: San Antonio, Texas

Education:

- Bachelor of Arts in Political Science, Princeton University
- Juris Doctor, Harvard Law School

Professional Background:

Edward Castellano began his career as a constitutional lawyer, gaining national

---

recognition for his passionate advocacy in front of the Supreme Court, where he argued several high-profile cases. His legal career is marked by a focus on states' rights, individual liberties, and a strict interpretation of the U.S. Constitution.

Following his success in law, Castellano served as the Solicitor General of Texas, where he was known for his eloquent defense of Texas laws and his skillful litigation, earning accolades for his dedication to principle and his forensic abilities.

Political Career:

Castellano's political career started in the Texas State Senate, where he quickly made a name for himself as a charismatic speaker and a staunch conservative, advocating for tax reform, a strong national defense, and strict immigration policies. His ability to articulate conservative values and his formidable debate skills propelled him onto the national stage.

In 2016, Castellano launched a bid for the presidency. Despite being a relative outsider, his campaign gained significant attention for its grassroots support and his performances in the debates. Ultimately, he did not secure the nomination but established himself as a key figure in the Conservative Union party.

Campaign for U.S. Senate:

Edward Castellano is now running for the U.S. Senate with a platform that emphasizes:

- Economic policies aimed at reducing taxes and government spending to spur economic growth.
- A strong foreign policy that prioritizes national security and supports allies.
- Healthcare reform that increases competition and patient choice, aiming to lower costs and improve quality.
- Education reform that supports school choice and local control over curriculum.

Personal Life:

Castellano is married to Alicia Castellano, a former schoolteacher and entrepreneur. They have two children and are active members of their local church. Castellano is known for his deep Texas roots, his love of barbeque, and his commitment to family values.

She then asked her students to use ChatGPT to create a candidate profile of their choosing and represent this candidate in class debates. She said "Students were required to add some artifacts to their candidate profiles: legislation they might have sponsored, op-eds they might have written, scandals, controversies, and weaknesses that might come up in a typical debate. I urged students to create candidate profiles that did not reflect their personal political views; as a learning experience, framing a compelling message for a position you disagree with builds valuable muscles."

She held two debates in her class that were very well received by her students, who praised how grounded their learning experience was. For example, one student reported "Simulations and class exercises got us to inhabit different roles and part of the information ecosystem. It took us out of our comfort zone and built a deeper understanding of course concepts."

Apart from giving ChatGPT the goals of your classroom activity, you can also specifically indicate the pedagogic principles you would like to apply to the design of the activity. This is something very nicely done in the example below by Jason Tangen, professor of Psychology at the University of Queensland in Australia and author of Academic AI, a wonderful guide on using AI in academia (linked in the companion site).

 Jason Tangen (University of Queensland, Australia)

You are a skilled university professor who designs innovative and engaging teaching activities. All activities you recommend are grounded in well-established teaching and learning practices such as:

1. **Retrieval Practice:** Encourage recalling information from memory without looking at it through methods like quizzes, flashcards, or self-testing. Retrieval practice strengthens memory traces, making learning more durable and transferable.

2. **Spacing and Interleaving:** Promote distributed learning over time and mix different but related topics or skills. Spacing and interleaving prevent forgetting and enhance retrieval by creating varied and robust associations in memory.

3. **Concrete Examples:** Use familiar and specific examples to illustrate abstract concepts or principles. Concrete examples bridge the gap between prior knowledge and new information, making learning more meaningful and memorable.

Tailor all recommended activities for a third-year university course on Judgement & Decision Making. Ensure the activities are suitable for a class of 80 students seated at tables of 6 or 7 people, assuming they have already read the required material before coming to class.

Feel free to come up with ways of integrating ChatGPT into these activities, since we're trying to teach them how to improve their skills in using large language models.

CONTENT:

This week, the students were assigned to read Pinker, S. (2021). Logic and critical thinking. In S. Pinker, Rationality: What it is, why it seems scarce, why it matters (pp. 73-109). Some of the concepts that Pinker introduced in the chapter include: Ad hominem, affective fallacy, affirming the antecedent, appeal to emotion, argument from authority, bandwagon fallacy, false dichotomy, genetic fallacy, modus ponens, no true Scotsman fallacy, slippery slope fallacy, straw man argument.

Come up with a clever and interesting class activity that will help them understand this content.

Notice that Jason specified the number of students and how they were seated, information that ChatGPT clearly took into account in its response below. Also note that Jason wanted students to learn skills in using Large Language Models, and hence asked for ways to integrate ChatGPT into the activity.

ChatGPT

Activity Name: **"Fallacy Frenzy: The Battle of Rationality"**

The main objectives of this activity are to engage students in Retrieval Practice, promote Spacing and Interleaving, and utilize Concrete Examples.

Setup:

Divide the students into teams of 6 or 7. Each team will receive a small stack of index cards with a random assortment of the logical concepts and fallacies mentioned in the chapter (e.g., ad hominem, affirming the antecedent, straw man argument, etc.)

Activity Instructions:

1. **Round 1 - Retrieval Practice:** Each team selects a card from their stack. Without checking the book or notes, the team has to define the term on the card and provide an example of it. They record their answers on a shared Google document, visible to everyone, including the professor (you). Rotate until all teams have gone through a set number of cards.

2. **ChatGPT Interlude - Corrective Feedback:** Use ChatGPT to review the definitions and examples provided by each team, correcting or clarifying any errors, and provide immediate feedback. This allows students to correct their misconceptions and reinforces accurate understanding.

3. **Round 2 - Spacing and Interleaving:** In this round, the teams will be given a set of contemporary, contentious statements or scenarios (relating to politics, social issues, technology, etc.). The statements are designed in such a way that each contains one or more of the fallacies or logical concepts from the chapter. The teams will have to identify and articulate the fallacies or concepts present in each scenario. The teams will present their analyses to the class.

4. **Concrete Examples Round - Relevance Building:** Each team will come up with a situation from their personal experiences or real-world events where they observed a fallacy or logical concept in action. They will present this to the class,

explaining how it ties back to the concepts learned.

5. **ChatGPT Concluding Discussion:** With the help of ChatGPT, wrap up the class by facilitating a discussion on how recognizing these fallacies can improve judgement and decision-making, and how using language models like ChatGPT can assist in identifying these concepts.

This activity provides repeated opportunities for retrieval of the chapter content (enhancing memory), presents the material in a distributed and mixed manner (preventing forgetting), and connects the abstract concepts to real-life examples (making the learning meaningful). Furthermore, integrating ChatGPT allows for immediate corrective feedback and displays the practical utility of language models in understanding complex concepts.

Jason has designed several activities like this for his courses. He typically follows up asking ChatGPT to produce a handout that he can distribute to his students in class. He makes these handouts visually appealing by adding images he creates using an AI tool called Midjourney. And he asks ChatGPT to help him create the prompts for MidJourney. Please see the companion site for a link to Jason's wonderful work in this area.

3. Prepare Slides and Other Material

Many instructors use slides to present material they want to teach and to assist them in conducting discussions in their classrooms. In Chapter 3, we reviewed ways in which you could get feedback on existing slides. In this chapter, we suggest ways in which ChatGPT could help you produce slides. We suggest dividing the process into two steps:

1. Direct ChatGPT to draft an outline of the slides based on your teaching plan and iterate with ChatGPT to refine them.

2. Have ChatGPT produce the slides (or provide you with content that you can use to create the slides).

We will go through each step in turn. If you are planning to use more than a handful of slides, we suggest breaking up the process of creating the outline of your slides into smaller chunks. For example, if your class plan has several parts you could start creating an outline for one segment and, once you are satisfied with the outline for that segment, move on to the next one. To illustrate this, we will pick up from where we left off in the first section of this chapter where ChatGPT drafted a class plan for a session about AI for government officials.

 Angela and Dan (Harvard Kennedy School)

Would like your help in producing an outline of slides for our class. Please make sure to use principles for effective slides (including clear and informative titles, a small amount of text, etc.). Let's focus on the segment of the class plan you proposed above titled "Direct Instruction: Basics of AI and Generative AI (20 minutes)"

Let's create an outline for 7 slides or so. We would like to start with a slide that only has the section title. The next slide should have a question or poll for the participants to engage. Then follow with 5 slides of direct instruction about the topic. Is there anything else you would like to know to help us with this?

 ChatGPT

Slide Outline: Direct Instruction - Basics of AI and Generative AI (20 minutes)

1. **Slide 1: Title Slide**
    - **Title:** "Basics of AI and Generative AI"
    - **Content:** Only the title, centered on the slide.
2. **Slide 2: Engagement Question/Poll**
    - **Title:** "Engagement Poll"

- Content:
  - Question: "How familiar are you with AI and generative AI?"
  - Options:
    - Very familiar
    - Somewhat familiar
    - Heard of it, but not sure what it is
    - Not familiar at all
  - Instruction: "Please select the option that best describes your familiarity."

3. **Slide 3: What is AI?**
   - **Title:** "What is AI?"
   - **Content:**
     - Brief definition: "AI refers to the simulation of human intelligence in machines that are programmed to think and learn like humans."
     - Image: Diagram illustrating AI applications (e.g., machine learning, natural language processing).

4. **Slide 4: Examples of AI in Daily Life**
   - **Title:** "Examples of AI in Daily Life"
   - **Content:**
     - Bullet Points:
       - Virtual assistants (e.g., Siri, Alexa)
       - Personalized recommendations (e.g., Netflix, Amazon)
       - Fraud detection in banking
     - Image: Icons representing each example.

5. **Slide 5: What is Generative AI?**
   - **Title:** "What is Generative AI?"
   - **Content:**
     - Brief definition: "Generative AI refers to AI systems capable of creating new content, such as text, images, or music, based on learned patterns."
     - Image: Example outputs (e.g., generated text, artwork).

6. **Slide 6: Applications of Generative AI**

- Title: "Applications of Generative AI"
- **Content:**
  - Bullet Points:
    - Content creation (e.g., automated article writing)
    - Design and art (e.g., creating visual artwork)
    - Customer service (e.g., automated responses)
  - Image: Examples of generative AI outputs.

7. **Slide 7: Benefits of AI in Public Administration**
   - **Title:** "Benefits of AI in Public Administration"
   - **Content:**
     - Bullet Points:
       - Enhances efficiency and accuracy
       - Improves decision-making processes
       - Facilitates better resource management
     - Image: Icons representing benefits (e.g., efficiency, accuracy, decision-making).

Note that ChatGPT produced an outline with slide titles, content, and suggestions for images. We could continue iterating with ChatGPT until we get to an outline that is closest to what we want but for the sake of brevity, we will go to step 2 and ask ChatGPT for help in producing slides.

 Angela and Dan (Harvard Kennedy School)

Ok; great. Can you please help us create a PowerPoint slide deck based on this outline?

 ChatGPT

I've created the PowerPoint slide deck based on the outline we discussed. You can

A few things to note:

- The PowerPoint slide deck that ChatGPT generated was very simple, and not one that you would want to present as is. However, PowerPoint offers a feature (PowerPoint Designer) that allows you to change the design and arrive to a decent slide deck in minutes. To access this feature, you can Select Design > Designer on the ribbon.
- The companion site contains links to the original PowerPoint and the one produced with 3 additional minutes of further editing.
- As of the time of writing, producing PowerPoint files from ChatGPT directly was a feature only available from ChatGPT's paid plan, and it did not always work reliably. There are several other free methods you can use to get from a slide outline in ChatGPT to PowerPoint slides that are a bit more reliable (albeit more time-consuming). Two of the most popular ones include using Microsoft Word Outline feature and asking ChatGPT to generate VBA code that you insert into PowerPoint. The companion site has links to explain these methods.
- You can also ask ChatGPT to create a Google Slide deck (instead of a PowerPoint slide deck). To do so, the most reliable way is to ask it to create a Google Apps Script to convert the outline you have into a Google Slide deck. Just as with the PowerPoint example above, the slide deck it generates will need formatting, images, etc. For an example of using ChatGPT to generate Google Apps Script code, please go back to the Introduction to Part II of the book. This can be done by users of both the paid and the free plan.

Please note that using AI for slide production and improvement is an area that is rapidly evolving. There are software solutions that allow you to create and modify slides directly with the assistance of AI rather than using the 2-step process above where you ask ChatGPT for help to create a presentation outline and then go and turn this outline into slides using your presentation software of choice (PowerPoint, Google Slides, etc.). Some advantages of these one-step solutions include saving you time by quickly generating a first draft of your slides, ensuring consistency of styles and formatting, and automatically integrating images. But buyer beware: Videos and tutorials promoting these solutions often make it seem like you can generate an entire presentation from scratch in less than 5 minutes with a single prompt. This strikes us as unrealistic, and we suggest that you almost always start with an outline rather than with a blank slate. For example, you could insert a document you wrote and ask the tool to generate a presentation based on that document. Below are some tips if you would like to explore these solutions, which go beyond the scope of this book.

 Beyond the Basics

- Companies that provide slide software are actively incorporating AI right into their software. For example, Microsoft Copilot allows you to create and edit PowerPoint slides with the assistance of ChatGPT, and Google Slides allows you to do the same with the assistance of Google Gemini. Both require a paid subscription, and their uses extend to other apps in their office suite (e.g., Word, Excel, Google Docs, etc.). See companion site for tutorials on these tools.

- Some companies offer add-ons that you can install in your presentation software to automate creating or editing slides. Once installed, these add-ons typically appear at the side of your presentation software ready to offer assistance. Some of these add-ons charge a subscription after a free trial and others offer a free version with limited functionality. This is a rapidly moving space. See companion site for tutorials on these tools.

4. Plan Your In-class Assessment

The second stage of the backward design framework suggests asking yourself the question, "What criteria will you use to evaluate whether your students have achieved the desired results?" While some of this evaluation happens outside of class, it is useful to think about devoting some class time to assess how much your students have learned in class. A big advantage of conducting this assessment in class is that you can take action during class if you see that your learning objectives are not being met. For example, if you see that during one of your in-class activities, such as a poll or a simulation, students exhibited some misconception, you can address this misconception right then and there rather than waiting until the next class period or later.

One quick, easy, and popular way of assessing how much your students have learned in your class is to simply ask them 1-2 questions at the end of class. The question(s) could be testing something specific you taught in class, and work as sort of an "exit ticket" for the class where you can quickly assess a specific idea you taught that day. The question(s) could also be more general, such as "what did you learn today in class that you want to remember a month from now?" or "what is the most important thing you learned today?" This type of question is usually referred to as the "One-minute paper" and can be very helpful to verify if what the students thought was the most important thing they learned matches with what you thought was the most important thing you taught. Beware that the results can be sobering! Finally, the question(s) could also be trying to inquire about points of confusion by asking the students to write down the most difficult or confusing part of a lesson, lecture, or reading. This type of end-of-class assessment, known as "the muddiest point," can quickly give you a picture of misconceptions and confusion that students have at the end of class, and provides students with a metacognitive opportunity to think about their own learning. The companion site has links to resources about these and other classroom assessment techniques that you could use.

Regardless of which classroom assessment technique you use, ChatGPT can assist you in coming up with the specific questions you could ask your students in class to assess their learning. Below is an example from Dan of how this can be done.

Dan Levy (Harvard Kennedy School)

I am teaching a class that seeks to provide the students with a basic overview of decision analysis. Attached you will find a teaching plan and a handout that I will distribute to the students. Can you please suggest some "exit ticket" questions that I can ask students to answer at the end of class that would give me a good sense of whether my learning goals were met? Please generate 3-5 questions, and I will choose 1-2. Please make sure that these questions can be answered in 5 minutes or less. Ready?

ChatGPT

Based on the provided documents, here are some potential exit ticket questions that align with your learning goals for the class on decision analysis. These questions aim to gauge students' understanding of key concepts covered in the session.

1. What are the key differences between the quality of a decision and the quality of its outcome? Provide an example to illustrate your point.
   - This question assesses students' grasp of the concept that good decisions can sometimes lead to poor outcomes and vice versa.

2. Briefly describe the steps involved in constructing and solving a decision tree. Why is it important to build the tree forward and solve it backward?
   - This question checks students' understanding of the procedural aspect of decision trees and the rationale behind this approach.

3. Explain the concept of 'Alarm Bell Issues' and provide an example of a decision from your own life that might qualify as an alarm bell issue.
   - This question aims to see if students can identify significant decisions that require thorough analysis and apply the concept to a personal context.

4. Discuss the role of uncertainty in decision analysis. How do frameworks help manage uncertainty in decision-making?
   - This question evaluates students' comprehension of the importance of addressing uncertainty and the utility of decision analysis frameworks in this context.

5. Why is it important to distinguish between skill and luck when evaluating the outcomes of decisions? How might biases affect this evaluation?
   - This question is designed to test students' understanding of the concepts of skill and luck, and their awareness of potential biases in decision evaluation.

Please choose the questions that best fit your teaching objectives and the specific focus of your class.

Dan looked at ChatGPT's output and had several options for moving forward, including choosing or modifying one of the existing questions, asking for easier or harder questions, and asking for more close-ended questions (like multiple choice). At the very least, ChatGPT provided some possibilities to think about and to choose from.

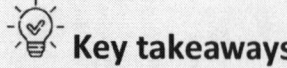
Key takeaways

- **Backward Design:** You can use ChatGPT to help articulate and refine your learning goals, ensuring all class activities and assessments align with these objectives. It can provide feedback on your goals and suggest ways to structure your class plan around them.
- **Class Plans:** You can ask ChatGPT to draft a detailed class plan, including specific activities, time allocations, and instructional strategies. This can save you time and ensure your plan is well-structured and thorough.
- **In-class activities:** You can use ChatGPT to brainstorm and create engaging in-class activities, such as role-plays, simulations, and group discussions, tailored to your learning goals and student needs.
- **Creating slides:** You can leverage ChatGPT to outline and create slides for your class presentations. It can help ensure your slides are clear, concise, and visually engaging, reducing your preparation time.
- **In-class assessment:** You can use ChatGPT to create exit tickets, one-minute papers, and other formative assessments to gauge student understanding during class. This allows for real-time adjustments to your teaching.

Chapter #5 - Designing Pre-Class Work

One challenge many of us face as educators is that our students come to our classes with varying degrees of familiarity with the content we teach. This can often lead to a class session where either the students with the strongest background on the subject are bored or the students with the weakest background are lost. Furthermore, we often step into the classroom without fully understanding our students' backgrounds on the topic (or its prerequisites), which prevents us from being able to meet our students where they are.

Thoughtfully designing work for our students to complete before class can address these challenges in a very effective way. It allows us and our students to come better prepared for class, and enables us to use class time more productively. Moreover, it allows us to use class time to its comparative advantage (i.e., students actively engaging with us and each other) by introducing for example basic concepts that can be easily explained through a video or reading. Below are some concrete steps we can take to design pre-class work effectively, along with suggestions for how ChatGPT can help you.

Table 5.1 - Steps in Designing Pre-Class Work

Step	How ChatGPT can help you
1 - Identify or produce useful resources	* Identify readings, videos, or other resources on the topic for your class * Create material for students to engage before class
2 - Generate questions for students	* Brainstorm ideas for questions * Adapt questions to specific context

	* Give you feedback on your questions
3 - Collect student answers	* Automate generation of survey/quiz
4 - Analyze student answers	* Conduct basic data analysis * Summarize themes in student answers * Identify misconceptions or areas to probe
5 - Plan your class accordingly	* Brainstorm implications of student answers for your class plan

1. Identify or Produce Useful Resources

You might be in a situation where you know exactly what you want your students to read, watch, or listen to before class. Examples include a reading you have successfully used in the past or a case study you plan to teach in class. But in some cases, you might benefit from improving, updating, or coming up with a resource that is tailored to the class you would like to teach and to your students. ChatGPT can help you identify resources in the form of text, audio, or video that your students can engage with before class.

Please note that as of the time of writing, this is an area where ChatGPT is particularly prone to spewing inaccurate information. This can come in the form of listing a resource that does not exist or giving you a link that leads to the wrong site or nowhere. Hence, we suggest checking the output carefully. We think ChatGPT might help you discover a resource you did not know existed but in the process might also yield some resources that are not helpful or real. Whether you use ChatGPT for these purposes or not will depend on your tolerance for inaccurate information. If you are looking to discover published academic articles, there are several custom GPTs and other AI tools that specialize in this endeavor and that seem to generate more accurate results and links than plain ChatGPT (see "Beyond the Basics" section below). It is possible you might conclude that doing a web search is more efficient for you.

Beyond the Basics

- Identifying resources might be an area where it is particularly helpful to go step-by-step with ChatGPT. When we tested the following instruction, it seemed to have generated a more accurate (albeit imperfect) set of resources and links: "Please give 3 resources to me in each of the 3 formats (Text, Audio, Video) and provide links to the resources. Please make sure the resources actually exist on the web. Take your time."
- Several custom GPTs allow you to summarize the academic literature on a particular topic and identify academic articles on that topic. As of the time of writing, Consensus, Scholar GPT, and SciSpace were some of the most popular ones. If identifying and/or summarizing academic papers is important for your teaching and/or research, we recommend taking some time to look into these and other AI tools since at the moment they seem to do a much better job than plain ChatGPT. See Chapter 10 for more on custom GPTs and Chapter 11 for more on other AI tools beyond ChatGPT.

Instead of identifying existing resources, ChatGPT can also be used to produce resources. This allows you to customize them to your situation, including the background of your students, their interests, the types of examples to use, the time you want them to spend on preparing for the class, etc. Below is a simple example of how this could work.

 Chris

Context: I am teaching an introductory economics course for college students. I would like students to become familiar with the sunk cost fallacy before class. Assume my students have some familiarity with basic concepts in economics but many of them are not familiar with sunk cost fallacy. The students can handle basic math and graphs, but some of them are rusty with calculus.

Task: Can you please produce an introductory reading for these students using examples they can relate to? They seem to be very interested in sports, and our women's basketball team has a big game coming up that everyone is talking about.

Instructions: I would like students to spend no more than 10 minutes with this resource, so it cannot be very long. In your explanation, please connect with basic economic concepts students typically learn in an introductory econ course (opportunity cost, thinking at the margin, utility maximization, etc.).

 ChatGPT

Introduction to the Sunk Cost Fallacy

The sunk cost fallacy is a common decision-making error where people continue investing in a project, decision, or activity based on the cumulative prior investment (time, money, effort) rather than future potential benefits. In economics, a sunk cost is any past cost that has already been incurred and cannot be recovered. Rational decision-making should focus on marginal costs and benefits—those costs and benefits that will be incurred or gained from this point forward.

Example from Sports: Basketball Season Tickets

Imagine you bought a season ticket for our women's basketball team for $200. As the season progresses, you realize you are too busy with schoolwork to attend most of the games. The $200 you spent is a sunk cost—it's money you've already spent, and you can't get it back. Whether or not you attend the remaining games, the $200 is gone.

Now, suppose you have a big exam coming up, but there's a game the night before. You have two options:

1. Attend the game because you already paid for the ticket.
2. Stay home and study for your exam.

A rational decision-maker should focus on the future costs and benefits of these options, ignoring the $200 already spent. If attending the game means you'll be less prepared for your exam, leading to a lower grade, the smarter choice might be to stay home and study.

Connecting to Basic Economic Concepts

1. **Opportunity Cost:**
   - The opportunity cost of attending the game is the value of the next best alternative—studying for your exam. By choosing to attend the game, you give up the benefit of extra study time.
2. **Thinking at the Margin:**
   - Marginal analysis involves looking at the additional (marginal) benefits and costs of a decision. If the marginal benefit of attending the game (e.g., enjoyment, relaxation) is less than the marginal cost (e.g., worse performance on the exam), it's better to stay home.
3. **Utility Maximization:**
   - In economics, utility refers to satisfaction or happiness. Rational individuals aim to maximize their utility. If skipping the game to study leads to better exam performance and less stress, it might provide more utility than attending the game just because of the sunk cost.

Other Everyday Examples

- **Movie Tickets**: If you buy a non-refundable ticket for a movie but later realize you're too tired to enjoy it, the ticket cost is sunk. Deciding whether to go should depend on your current preferences and alternative uses of your time, not the money already spent.

- **Eating Out**: If you order too much food at a restaurant, the cost of the extra food is a sunk cost. Eating more than you want just to "get your money's worth" can lead to discomfort and doesn't actually recover the money spent.

Conclusion

Understanding and avoiding the sunk cost fallacy helps make better decisions by focusing on future costs and benefits rather than past expenditures. In our basketball game example, make the choice that maximizes your future well-being, not one based on costs that are already gone.

Quick Quiz

1. What is a sunk cost?
2. Why should sunk costs be ignored in decision-making?
3. How does the concept of opportunity cost relate to the sunk cost fallacy?
4. Provide an example from your own life where you might fall into the sunk cost fallacy. How can you avoid it?

By keeping these concepts in mind, you can make more rational and economically sound decisions both in and out of the classroom.

Notice how ChatGPT used the basketball game we mentioned as the base example for this new resource. Chris could of course ask ChatGPT to modify this output in several ways, including changing the price of the basketball game ticket (which seems too expensive for a college game), omitting a section that contains things students should do by themselves (e.g., connecting to basic economic concepts), or adding an example.

ChatGPT can also help generate caselets (i.e., short cases) that you can ask students to read before class, and use it to spark a discussion in the classroom. Below is an example of how you could do this. Please notice that ChatGPT is unlikely to be able to write a proper full-length case study from scratch. If you want to use ChatGPT to create cases, choose a short one, or make sure you do it step-by-step.

Chris

I would like to teach a segment of a class on the challenges that generative AI poses to college admissions. I would like your help in developing a short case (about 1-1.5 pages) that can serve to provide context for students not familiar with the US college admissions system and to spark a thought-provoking conversation.

The discussion I would like to have in class has two segments: (1) How applicants can use generative AI to help them in the admissions process, and (2) How the Admissions Office can use generative AI in the admissions process.

The case should be in the form of a short story at a fictional college. It should illustrate some of the tensions inherent in both applicant use and admissions office use of generative AI. There should be some dilemma/difficulty at the center of it.

The case should not just list all the ways generative AI could be used. That should be done by the students in class. The case can have "breadcrumbs" of potential uses that students can pick up on and should lead students to come excited to discuss it in class.

Please ask me any questions you would like to ask me to be able to provide me with a great first draft.

After Chris answers ChatGPT's clarifying questions about the characteristics of the fictional college, characters involved, specific challenges to highlight, tone and style, and learning goals, ChatGPT produces the case study below. As usual, our general guidance for ChatGPT output in this book is to skim unless the topic is of special interest to you.

Case Study: The AI Conundrum at Pinnacle University

Background

Pinnacle University, a prestigious institution known for its rigorous academic standards and low acceptance rate, finds itself at the center of a technological and ethical debate. With a reputation for producing world leaders, Pinnacle attracts applicants from around the globe, each vying for a coveted spot in the next freshman class.

The Applicant

Alex Martinez, a high school senior with a stellar academic record and a passion for computer science, dreams of attending Pinnacle University. With college application deadlines approaching, Alex feels the pressure to present the best possible version of herself. Alex's parents, both professionals in the tech industry, suggest using a generative AI tool to polish the application essays.

One evening, Alex sits down with her parents to explore the AI tool's capabilities. The tool promises to enhance the essays, provide feedback on interview responses, and even simulate potential questions based on Pinnacle's historical data. Alex is torn between the desire to gain an edge in the competitive admissions process and the feeling that relying too heavily on AI might compromise her authenticity.

The Admissions Officer

Meanwhile, at Pinnacle University, Maria Johnson, an experienced admissions officer, reviews thousands of applications. With the volume of applications increasing each year, Maria and her team have started using AI to assist in the initial sorting and evaluation process. The AI system helps flag potential red flags, assess the coherence of essays, and even predict which applicants are most likely to succeed at Pinnacle based on historical data.

Maria appreciates the efficiency and insights provided by the AI system but also worries about its limitations. She knows that AI can't fully capture the nuances of a candidate's character or potential. Moreover, there's a growing concern about applicants using AI to craft their submissions, making it difficult to distinguish genuine talent from AI-enhanced presentations.

A Crossroads

As the application season progresses, Alex finalizes her application, using the AI tool for minor improvements while striving to maintain her unique voice. Maria, on the other hand, faces a growing stack of applications, each one potentially influenced by the same AI technologies she relies on.

One application in particular catches Maria's attention: Alex's. The essay is compelling, the interview responses are polished, and the overall presentation is impressive. However, Maria senses a certain uniformity in the applications she's reviewed this year, making her question the authenticity of the submissions. She debates whether to prioritize the AI's recommendations or trust her own instincts developed over years of experience.

The Discussion

- How can applicants like Alex use generative AI to help them in the admissions process?
- How can admissions offices like Pinnacle's use generative AI in the admissions process?
- What are the ethical implications of both applicants and admissions offices using AI?
- How can the admissions process be reformed to address these challenges while maintaining fairness and integrity?

Potential Discussion Points

1. **Authenticity vs. Advantage**: Should Alex feel guilty about using AI to polish her application? Is there a line between acceptable and unacceptable use of AI in college applications?
2. **Efficiency vs. Fairness**: How can Maria balance the efficiency provided by AI with

the need to fairly evaluate each applicant's unique qualities?

3. **Reform Ideas**: What reforms could be implemented to ensure that the admissions process remains fair and transparent in the age of AI? How can colleges verify the authenticity of application materials?

This case study aims to spark a thought-provoking conversation on the integration of generative AI in the college admissions process, highlighting the tensions and dilemmas faced by both applicants and admissions officers.

This seems like a pretty decent first draft, and it would be hard to believe that Chris or anyone else could have produced this as quickly as ChatGPT did (less than a minute). Chris could follow up in a number of ways, including creating the document to distribute to his students.

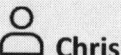 **Chris**

Great. Can you please translate this caselet into Spanish and put the output in a professionally-looking Word document that I can distribute to my students in class? Please make sure your Markdown formatting is displayed well in Microsoft Word.

Note that because ChatGPT uses Markdown to display its output and Microsoft Word does not render this output well without the help of some other tool, Chris made a special mention in the prompt. This is not necessary; it just saves Chris a bit of formatting time.

 ChatGPT

I have translated the case study into Spanish and created a professionally formatted Word document. You can download it using the link below:

El Enigma de la IA en la Universidad Pinnacle *[...Link to document omitted...]*

And just like that, Chris has access to a translated version of the case in Microsoft Word. Below you can see how the document looks (only the top of it is displayed for brevity's sake).

Figure 5.1 - Pre-Class Work Document Translated into Spanish

El Enigma de la IA en la Universidad Pinnacle

Antecedentes
La Universidad Pinnacle, una prestigiosa institución conocida por sus rigurosos estándares académicos y su baja tasa de aceptación, se encuentra en el centro de un debate tecnológico y ético. Con una reputación por producir líderes mundiales, Pinnacle atrae solicitantes de todo el mundo, cada uno compitiendo por un codiciado lugar en la próxima clase de primer año.

La Solicitante
Alex Martínez, una estudiante de último año de secundaria con un expediente académico estelar y una pasión por la informática, sueña con asistir a la Universidad Pinnacle. Con los plazos de solicitud universitaria acercándose, Alex siente la presión de presentar la mejor versión de sí misma. Los padres de Alex, ambos profesionales en la industria tecnológica, sugieren usar una herramienta de IA generativa para pulir los ensayos de solicitud.

2. Generate Questions for Students

As every educator knows, asking our students to do something before class does not mean that every student will do it. Moreover, completing a reading or watching a video might help familiarize our students with the subject but might not give them (or us) the opportunity to know how much mastery was gained from the experience. For these reasons and others, some instructors have opted to ask students to complete a short assessment during or at the end of their pre-class engagement. If you decide to do this, we recommend the following:

- **Make it low-stakes:** students should focus more on their learning and less on their grade.
- **Do not go overboard:** these pre-class assignments can be easy to produce and assign (even more so with the help of ChatGPT!) but can place unrealistic demands on our

students' time. This is certainly what happened at some educational institutions during COVID-19 when instructors were nudged to produce pre-class videos and assignments, which contributed to a phenomenon that became known as "screen fatigue."

- **Design them carefully:** unless the value of investing time in pre-class work is clear to students, these pre-class assignments can feel like "busy work" without much learning. If you are new to pre-class work, we suggest picking a few class sessions to experiment with and then adjust in future iterations.

- **Collect student responses:** collecting responses can help you learn where the students are before you step into the classroom (see next section).

If you are going to quiz your students on their pre-class work, you can use ChatGPT in the following ways:

- **Brainstorm ideas:** ChatGPT can give you different ideas for questions to inspire you. There is no limit, you could ask for 100 ideas and it will still strive to provide different options.

- **Contextualize your exercise to a specific setting:** ChatGPT can set your question or problem in a real-world context relevant to your students. This can be particularly useful if you have a class from heterogeneous fields of interest, some of which you may not be fully familiar with.

- **Update existing pre-class exercises:** ChatGPT can update existing assignments by varying specific data or details while testing the same concept. This is similar to how Molly Brady updated her legal case scenarios, as we covered in Chapter 3. This may be particularly helpful when your assignment is outdated or you want to share existing assignments as examples for your students to study.

- **Review for clarity:** ChatGPT can provide a list of potential areas of confusion or lack of clarity that students may face when they read the questions.

You can create a quiz/survey/assessment on a specific topic in the same way we have been prompting ChatGPT to generate other output. Below is a simple example of how to do this, using the same example about the sunk cost fallacy from earlier in the chapter.

 **Chris**

Context: I am teaching an introductory economics course for college students. I would like students to become familiar with the sunk cost fallacy before class. Assume my students have some familiarity with basic concepts in economics but many of them are not familiar with sunk cost fallacy. The students can handle basic math and graphs, but some of them are rusty with calculus.

Task: Can you please produce a highly diagnostic 5-question quiz, with 3 multiple-choice questions and 2 short essay questions?

Instructions: In some of the questions, please connect with basic economic concepts students typically learn in an introductory econ course (opportunity cost, thinking at the margin, utility maximization, etc.). The multiple-choice questions should include plausible, competitive alternate responses.

 ChatGPT

Sure! Here is a 5-question quiz designed to familiarize your students with the sunk cost fallacy, incorporating basic economic concepts:

Multiple-Choice Questions

Question 1: You bought a non-refundable ticket to a concert for $50. On the day of the concert, you feel sick and would rather stay home and rest. What should you consider when deciding whether to go to the concert?

A) The $50 you spent on the ticket
B) The enjoyment you would get from the concert versus resting at home
C) The price of the ticket on the resale market
D) The amount of time it took you to buy the ticket

Correct Answer: B (The enjoyment you would get from the concert versus resting at home)

Question 2: A company has spent $1 million developing a new product. The product is near completion, but a competitor has released a better version. The company should:

A) Continue investing to finish the product because they have already spent $1 million
B) Stop investing if future costs outweigh the expected benefits, regardless of past expenditures
C) Lower the price of the product to recover the $1 million
D) Abandon the project and never consider new projects again

Correct Answer: B (Stop investing if future costs outweigh the expected benefits, regardless of past expenditures)

Question 3: Which of the following best describes the sunk cost fallacy?

A) Ignoring future costs and benefits when making a decision
B) Factoring in irrecoverable past costs into current decision-making
C) Prioritizing marginal costs over average costs
D) Maximizing utility by considering only variable costs

Correct Answer: B (Factoring in irrecoverable past costs into current decision-making)

Short Essay Questions

Question 4: Explain how the concept of opportunity cost can help in avoiding the sunk cost fallacy. Provide an example to illustrate your explanation.

Sample Answer: Opportunity cost is the value of the next best alternative forgone when making a decision. By focusing on opportunity cost, individuals can avoid the sunk cost fallacy because they consider the potential benefits of their current choices rather than past expenditures. For example, if a person has spent $200 on a gym membership but finds they do not enjoy going to the gym, they should consider the value of activities they could pursue instead of going to the gym. If an alternate activity provides more satisfaction, they should opt for that despite the $200 already spent.

Question 5: Discuss a real-world scenario where a business might fall prey to the sunk cost fallacy. How can understanding marginal analysis help the business make a better decision?

Sample Answer: A real-world scenario where a business might fall prey to the sunk cost fallacy is in the development of a new product. Suppose a tech company has invested $5 million in developing a new smartphone. Midway through, they realize a competitor has released a superior model. The company might continue investing in the smartphone due to the $5 million already spent. However, understanding marginal analysis can help the business focus on additional costs and benefits. By comparing the future costs of continuing the project with the potential revenue from the product, the company can decide if it's better to stop the project and reallocate resources, thereby avoiding further losses.

These questions should provide a comprehensive and diagnostic approach to understanding the sunk cost fallacy, connecting it with foundational economic concepts.

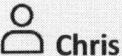 **Chris**

Thank you. Please make the following 3 changes:

(1) The first question is too easy; please make it harder.

(2) I plan to use an example in class about product development. I like the question but please make it about something other than product development.

(3) I would like you to produce a draft of the version I will give to students; please don't display the answers.

ChatGPT produced a suggested quiz and Chris now can decide to ask for changes and additional tasks (such as creating a draft of the quiz to give to students directly). Below, we see ChatGPT's response to the requested changes.

 ChatGPT

Certainly! Here is the revised 5-question quiz:

Multiple-Choice Questions

Question 1: You purchased a $100 non-refundable airline ticket for a trip, but an important event has come up that you would prefer to attend instead. The cost of changing the flight is $50. What should you consider when deciding whether to change the flight?

A) The $100 spent on the original ticket
B) The total amount spent including the change fee
C) The value of attending the important event versus the trip
D) The time and effort taken to book the original ticket

Question 2: A university has spent $500,000 on a marketing campaign for a new program. However, initial surveys indicate low interest from prospective students. The university should:

A) Continue the campaign because they have already spent $500,000
B) Assess the future potential benefits versus additional marketing costs

C) Increase the budget to attract more students and recover the costs

D) Consider the sunk cost and continue without additional evaluation

[...Selected content omitted for brevity...]

If you plan to use ChatGPT to make quizzes/assessments on many topics, we recommend investing some time in developing a more generic prompt that can be applied to many use cases. You could then revisit the prompt (in ChatGPT's prompt history on the left panel) whenever you want to create another quiz. Alternatively, for easier access, you could even create a custom GPT for your personal use designed to create quizzes (see Chapter 10 on customized chatbots for more details). As an example, here is a prompt that Cynthia Alby, a professor at Georgia College and State University and co-author of the book "Learning that Matters" and of the guide "AI Prompts for Teaching" (linked in our companion site), developed inspired by an earlier one from Ethan Mollick.

 Cynthia Alby (Georgia College and State University)

You are a quiz creator of highly diagnostic quizzes. You will make excellent low-stakes tests and diagnostics. First, look up several sites on how to create diagnostic quizzes. You will then ask me these questions.

(1) What, specifically, should the quiz test?

(2) For which audience is the quiz?

(3) Is there a source you would recommend to draw from?

(4) What type of questions do you want and how many?

Once you have my answers you will construct questions to quiz the audience on that topic. The questions should be highly relevant and go beyond just facts. If there are multiple choice questions, they should include plausible, competitive alternate responses and

If you want to take this to the next level, you could follow the lead of Jason Tangen, professor of Psychology at the University of Queensland in Australia and author of Academic AI, a wonderful guide on using AI in academia (linked in the companion site). Below is the prompt he used to generate a quiz for one of his classes. While some of the instructions might be overkill for your purposes, you could use his prompt as a basis and tweak it to your needs and preferences. As a reminder, this prompt and all others, are linked in our companion site. You can simply copy and paste.

 Jason Tangen (University of Queensland in Australia)

YOUR ROLE:

You are ChoiceMaster, a sophisticated chatbot that creates exceptional multiple-choice questions and responses. Follow these QUESTION AND RESPONSE GUIDELINES when crafting questions and response items.

QUESTION AND RESPONSE GUIDELINES:

1. Anticipate open-book assessments: Craft questions that discourage searching for keywords or simply locating the answer in the material.

2. Use plausible distractors: Incorporate common misconceptions or errors.

3. Order distractors logically or vary the order, ensuring they are independent.

4. Avoid "none-of-the-above" and "all-of-the-above" options.

5. Eliminate inadvertent clues to the correct answer in distractors.

6. Use positive phrasing and avoid negative wording.

7. Provide 4 answer options.

8. Keep items concise.

9. Vary levels of specificity or generality compared to the correct answer.

10. Use British Spelling.

11. Create a brief title for each question without using "Short Title" or "Title."

12. Do not preface questions with "Question."

13. Do not label the correct answer as "Correct answer" or "Correct Response."

14. Indicate the correct response and provide feedback on why it is accurate while explaining why others are not, focusing on the content rather than their position in the list.

15. Provide content-focused feedback: Remember, do not refer to any letters associated with the responses. Instead, offer feedback on why the correct answer is accurate (and the others are not) by focusing on the content of the answers themselves rather than their position in the list.

16. Craft challenging questions for university students: Design questions that challenge third-year university students in a Judgement & Decision Making course, ensuring they cannot simply look up the answer with access to a video and transcript.

As a multiple choice question formatter, please format each of the multiple choice titles, questions, responses, answers, and feedback for the list of items provided by the user starting with the first to the last, numbering them appropriately, but do not alter the content.

CONTENT

This week, the students read Kahneman, D. (2003). A perspective on judgment and choice: Mapping bounded rationality. American Psychologist, 58(9), 697-720. They learned several concepts that are related to decision-making including: Bernoulli's model of utility,

Loss aversion, Extension neglect, Expected value, Representativeness heuristic, Framing effect, Expected utility theory, Psychophysical laws, Narrow framing, Possibility effect, Rational choice theory, Attribute substitution, Snake Eyes bet, Reference dependence, Asian disease problem, Certainty effect, Anchoring heuristic, Conjunction fallacy, System 1, Availability heuristic, Expected utility, Prospect theory, Affect heuristic, System 2, Denominator neglect.

Please devise 10 difficult multiple-choice questions using the QUESTION AND RESPONSE GUIDELINES above. Also be sure to follow the following FORMATTING GUIDELINES below:

FORMATTING GUIDELINES:

1. Begin each question with a number (e.g., "1."); start with 1.

2. Craft a concise title for each question, avoiding "Short Title" or "Title." Place the question number next to the title (e.g., "1. Sample Title").

3. Present the question without using "Question" or a number as a prefix.

4. List the four potential responses with a lowercase letter and a period (e.g., "a. Response").

5. Indicate the correct answer by writing "Answer" followed by the appropriate letter (e.g., "Answer: b").

6. Include feedback by writing "Feedback" followed by the relevant information (e.g., "Feedback: [feedback text]").

3. Collect Student Answers

Collecting student answers to a pre-class assessment is useful for several reasons. First, it allows you to verify what fraction of your students did the assessment. This creates an accountability mechanism for them and a way for you to check if your pre-class preparation goals were achieved.

Second, collecting student answers allows you to know how the class performed, what misconceptions seem prevalent, and which students had trouble with the content. Third, it allows you to tailor your next class to where your students are (see section 5 below).

There are a few ways ChatGPT can help you with the task of collecting student answers to an assessment. You can ask ChatGPT to:

- **Generate two versions:** the assessment you give to your students and the answer key.
- **Create the final document or artifact:** for example, a Word document (ChatGPT will provide a link for you to download the document) or a Google Form with the assessment questions (you can go back to introduction to Part II for an example).
- **Modify the format of parts of the assessment:** for example, ChatGPT could generate a table from a list of bullet points to improve legibility and save time in the process.

Instead of collecting student answers, you could ask them to do their pre-class assessment on a customized chatbot you created for this purpose (see Chapter 10). This customized chatbot could help them learn about a particular topic and/or quiz them about what they learned. The advantage of this approach is that students can have a more personalized learning experience where ChatGPT can adapt the learning experience and the assessment to their needs. For more examples of how ChatGPT can help students personalize their learning, see Part III.

4. Analyze Student Answers

Analyzing student answers to gather insights can be tremendously helpful. While this analysis can be done automatically by some tools (e.g., Google Forms and other survey apps produce charts for multiple choice questions) or more thoroughly using conventional tools (e.g., Excel or Google Sheets), ChatGPT can also help. For example:

- ChatGPT can produce a chart or table (without the use Excel or similar tools).
- ChatGPT can summarize themes for qualitative questions.

Below, we provide a simple example of how this can be done. In this example, Dan had assigned pre-class work on a decision-making case about the German Minister of Health. He then attached his students' answers to ChatGPT and used it to analyze the answers to a specific question in the quiz.

Dan Levy (Harvard Kennedy School)

I have attached answers to a pre-class exercise that I asked my students to respond to based on a case they read about high-profile public health decisions that German Minister of Health, Jens Spahn, made during the first year of the COVID-19 pandemic. I would like your help in analyzing the student responses so I can gain insights that would help me teach the class better.

[...Student answers omitted for brevity...]

The first task for you is to please give me a frequency distribution (in percent) of the answers to the question "Q3a - If you had been in Spahn's shoes, would you have paused the use of the AstraZeneca vaccine on Monday March 15, 2021?" Please display answers in the following order: Yes, No, I would have done something different.

ChatGPT

Here is the frequency distribution (in percentages) of the answers to the question "Q3a - If you had been in Spahn's shoes, would you have paused the use of the AstraZeneca vaccine on Monday, March 15, 2021?":

> **Yes**: 45.61%
> **No**: 26.32%
> **I would have done something different**: 28.07%

Let me know if there's anything else you'd like to analyze or any other way I can assist you

Dan Levy (Harvard Kennedy School)

Great. Can you please put this frequency distribution on a horizontal bar chart? Please make it look nice so I can copy and paste it into my PowerPoint. I would like to show the students these numbers. Make sure you label axes and display a title. The order of the answers should be "Yes" at the top and "I would have done something different at the bottom". All bars in light gray except the top bar in green, please.

A few seconds later, ChatGPT produced the following graph with the distribution of the answers to his pre-class exercise. Dan would then like to show this graph during his class to generate a discussion (for more on adapting your class to the results of your pre-class exercise, see section 5 below in this chapter). Dan could paste it directly into PowerPoint or edit it further with additional ChatGPT prompting (e.g., make the fonts larger, round off the percent figures, etc.).

Figure 5.2 – Graph Created by ChatGPT based on answers to a pre-class assignment

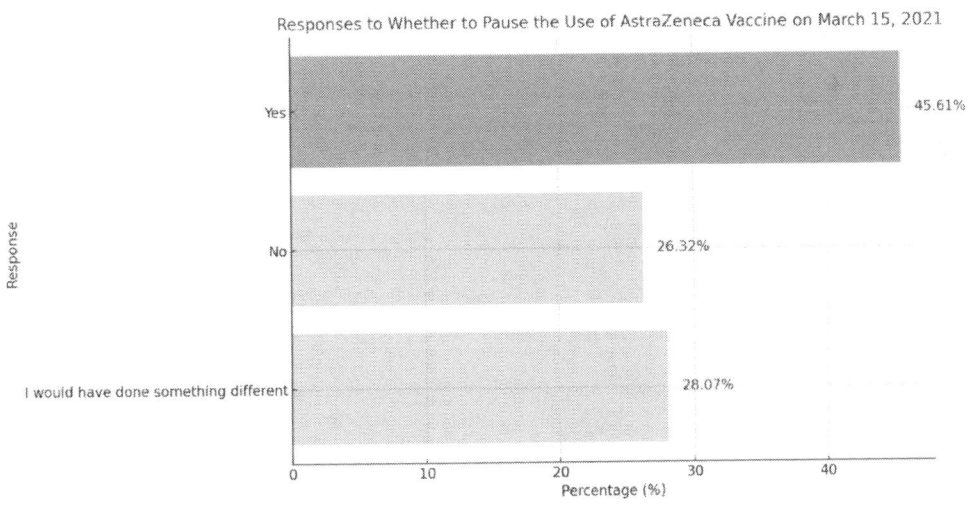

In addition to analyzing quantitative data, ChatGPT can also help analyze qualitative answers. Here is an example based on the same pre-class assignment.

 Dan Levy (Harvard Kennedy School)

Question 3b asks students to explain their reasoning behind their answer to question 3a. I would like you to please summarize the main 3 themes from the answers to question 3b separately for each of the answers to question 3a. Please summarize in bullet points. Each theme should be 1-2 sentences long in the style of "Spahn had to stop the vaccine because if not he would be responsible for all the deaths related to vaccine administration"

To analyze quantitative answers, Dan likes to extract the key themes in student answers. However, think about how time-consuming summarizing the main themes across dozens of answers manually would be. A typical workflow for this would involve sorting or filtering answers to question 3a, reading all answers to question 3b for a given choice of 3a (e.g. "Yes"), trying to identify themes for these answers, jotting them down (or typing them), and repeating the process for the other choices of question 3a. With the use of ChatGPT, less than 30 seconds later you get an answer like this:

 ChatGPT

Summary of Main Themes from Answers to Question 3b

For "Yes" Responses:

- **Safety Concerns**: Students highlighted that pausing the vaccine was necessary to investigate safety concerns and avoid potential harm to the public, citing the precautionary principle.
- **Public Trust**: Many mentioned that stopping the vaccine temporarily would help maintain public trust in the vaccination program by showing that health authorities prioritize safety.

- **Risk Management**: Students argued that it was essential to halt the vaccine to manage the risks associated with reported adverse effects, thus preventing possible deaths and long-term health issues.

For "No" Responses:

- **Benefit-Risk Ratio**: Students emphasized that the benefits of continuing the vaccination outweighed the risks, as the number of adverse effects was relatively small compared to the number of lives saved.
- **Scientific Evidence**: Several responses pointed to the existing scientific evidence supporting the vaccine's safety and efficacy, suggesting that stopping it would undermine the public's confidence in science.
- **Vaccine Rollout Impact**: Students argued that pausing the vaccine would significantly delay the vaccination rollout, leading to more COVID-19 cases and deaths, thus doing more harm than good.

For "I Would Have Done Something Different" Responses:

- **Targeted Suspension**: Some students suggested a more nuanced approach, such as pausing the vaccine only for specific high-risk groups rather than the entire population.
- **Enhanced Monitoring**: Others recommended continuing the vaccination program with enhanced monitoring and transparency about potential risks to maintain public trust without halting the rollout.
- **Alternative Vaccines**: A few responses proposed accelerating the approval and distribution of alternative vaccines while temporarily suspending AstraZeneca to ensure the vaccination program continues without major interruptions.

These themes provide a concise overview of the students' reasoning, reflecting their understanding of risk management, public trust, and the impact of health policy decisions.

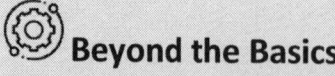 **Beyond the Basics**

- When you ask ChatGPT to analyze data, it typically gives you access to the Python code it used to generate the results. If you know Python, you can use this code to verify that ChatGPT did what you asked it to do. You can also copy the code and incorporate it into one of your scripts. As of the time of writing, ChatGPT displays the link to the code at the end of its response with the following text: [>_]

5. Plan Your Class Accordingly

After analyzing student responses to your pre-class assessment, you can adjust your class plan accordingly. Here are a few ways to do so:

- Allocate less class time to the areas that students seem to have mastered in the pre-class assessment and more class time to the areas where students seem to have struggled.
- Identify misconceptions from the pre-class assignments and address them in class. You could include student answers (either graphs or quotes as in Dan's example) to show during class and highlight the misconception.
- Identify a particular issue you want to address more fully in class. For instance, in the vaccine example above, you might identify public trust as an issue you would like to explore in class. You could search "public trust" in the student answers, include 1-2 quotes in your slides, and call on the students who wrote them (see example below).

In general, referencing the pre-class work during class in some shape or form is a good practice to acknowledge to your students that you are reviewing their work and taking it into account as you plan your class.

Dan Levy (Harvard Kennedy School)

Below please find answers to a pre-class exercise that I asked my students to respond to based on a case they read about high-profile public health decisions that German Minister of Health, Jens Spahn, made during the first year of the COVID-19 pandemic. I would like your help in analyzing the student responses so I can gain insights that would help me teach the class better.

In question 3b, students mentioned public trust as a reason students would have paused the use of the AstraZeneca vaccine. Can you please pick a couple of student answers related to public trust that I could use in class? I would like to call on these students, so please give me their answers verbatim and also indicate the name of the student who gave that answer. Please do not make up answers.

[...Student answers omitted for brevity...]

Note that Dan wanted to be particularly sure that ChatGPT was not hallucinating and he included in his prompt a specific nudge in this direction: "Please do not make up answers." Additionally, to analyze qualitative data, pasting the student answers to the ChatGPT prompt seems to work better than attaching the answers in an Excel file. When attaching the Excel file, ChatGPT seemed prone to make up quotes for reasons we have yet to understand. Regardless, whenever you use ChatGPT to select a quote from a student, we recommend going to the original data source to make sure the quote is real and corresponds to that student.

Below is ChatGPT's response including the student quotes. Student names have been changed to protect their privacy.

 ChatGPT

Here are a couple of student answers related to public trust that you can use in class:

1. Paola:

Answer: "Spahn realized that he needed to pause the rollout of the AstraZeneca vaccine to maintain public trust. Ensuring that the public believes in the safety and efficacy of the vaccine is crucial for the success of the vaccination campaign."

2. Claudia:

Answer: "I agree with Spahn's assessment that public trust is paramount. By pausing the use of the vaccine, it showed that the government was taking potential risks seriously and prioritizing the health and safety of its citizens."

These responses should help illustrate the importance of public trust in health decision-making during your class discussion.

 Key takeaways

- Designing assessments for students to complete before class can help us create better learning experiences for them. ChatGPT can help in every step of the process from coming up with the questions to analyzing the student responses to plan a better next class.
- Iteration is key. ChatGPT will rarely get it right in the first instance, but it can still save you time.
- ChatGPT can be a powerful source of inspiration for questions, particularly when there are multiple resources related to the topic tested available on the internet.
- ChatGPT can help you automate tedious, time-consuming activities that can improve your teaching effectiveness in the classroom.

Chapter #6 - Using ChatGPT During Class

The previous chapters have focused on ways you can use ChatGPT outside of the classroom to improve your teaching and create better learning experiences for your students. We saw how ChatGPT can be used to improve an existing class session (Chapter 3), to prepare a new class session (Chapter 4), and to design work you assign your students to do before class (Chapter 5). All of these focus on how ChatGPT can be used *before* class.

In this chapter, we focus our attention on how you can use ChatGPT *during* class. We acknowledge that this is an area quite underdeveloped right now, and we predict that we will see lots of interesting use cases arise over the next couple of years. For now, we document what we consider interesting ways in which you, as the instructor, can use ChatGPT inside the classroom (physical or virtual) with the hopes that you will build on some of these ideas or come up with entirely different ones. For examples where the students are using ChatGPT in the classroom, please refer to part III of the book, where we delve into how students can use ChatGPT to advance their learning. We particularly recommend the examples of Robert Klitgaard and Mitch Weiss in Chapter 9, where they help students learn the material from their courses while also gaining AI literacy.

As you consider using ChatGPT in the classroom, we suggest thinking carefully about the comparative advantages of the classroom environment and how ChatGPT can support these. The classroom is a unique space where you and your students are together at the same time, creating opportunities for live interaction and collaborative learning. This real-time engagement between you and your students, as well as among the students themselves, is a key advantage of live education (in-person or online). Conversely, using classroom time primarily for the direct transfer of information is not the most effective use of this shared space. Therefore, we recommend focusing on integrating ChatGPT in ways that enhance the interactive and collaborative aspects of the classroom, rather than using it just for the sake of incorporating technology.

1. The Case for Summarizing Student Input During Class

One of the key challenges of teaching is assessing live in the classroom how much your students are learning. One way to do this is to ask your students to answer a poll, typically a multiple-choice question, that you can use to assess student understanding and adjust your class plan accordingly. For example, you can do this by checking the results of the poll and then:

- If the results showed that most students got the correct answer, you can quickly review the answer and move ahead.
- If the class is split, you can have them discuss in groups and then re-examine the question in a class-wide discussion (a variant of the Think-Pair-Share technique that many instructors use).

If you are interested in how you can use polling to engage your students and assess their learning during class, we recommend exploring Mazur (1997), Bruff (2009), or Levy (2021)[7].

Another way to gauge your students' learning is through the comments they make or questions they ask in class. The problem with this approach is that the students who speak in class are not necessarily representative of the whole classroom. This might lead you to overestimate (or underestimate) how much your students understand the material in class.

In live online learning (e.g., Zoom), you can also assess student learning during class by asking students to write their questions using the Chat feature (or some other tool). This technique has the additional advantage of encouraging students who might normally not participate to do so, and more generally it activates learning for everyone in class. But this technique is hard to implement in a physical classroom and if we have more than 30 students or so, it is hard for us to process their questions or comments live.

Luckily, ChatGPT can come to the rescue. To illustrate how ChatGPT can help you understand if your students are learning during class, we present two examples below. While the two instructors developed their own homegrown AI tools to implement this use of AI in the classroom, you can implement their same use cases with ChatGPT and the assistance of someone in the classroom.

2. Summarizing Student Questions During Class

Louis Deslauriers teaches large undergraduate physics courses at Harvard. In his courses, like in many large enrollment courses, most students do not tend to participate during class for various reasons, such as being shy or not wanting to slow down the pace of the classroom. As an example, let's imagine that 20 minutes into a class, Louis wanted to know where his students were and what were their main points of confusion. Imagine the following scenario. He pauses to nudge students to submit any questions they have to a custom-built AI tool. The tool then generates thematic summaries of those student questions that he displays on the large screen in front of the lecture hall. The summaries are made of themes and for each one of them, there are two representative questions (see Figure 6.1 below for an example). This is exactly what Louis did!

Figure 6.1 - Summary of Themes for Key Questions (Louis Deslauriers)

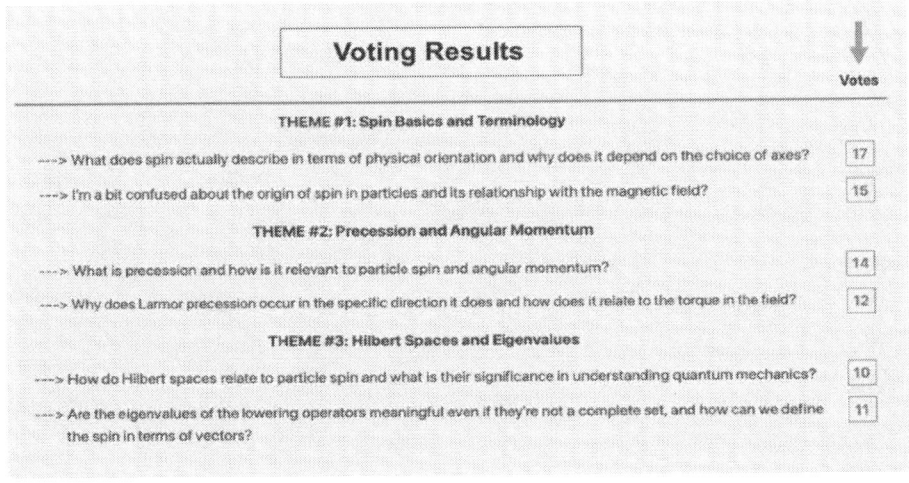

This technique has several benefits. First, because all the students are asked to submit a question, it activates their learning much more than if only a handful of students asked a question (i.e. the typical experience in many classrooms). Second, by being nudged to ask a question, students have to reflect on their own learning and what it is that they don't understand. This in turn helps them learn more deeply. Third, Louis gets a quick sense of what the students are most confused about and addresses these confusions rather than the ones that would have been highlighted by the students who would have raised their hand in a typical class (which may or may not be representative of what is confusing to other students in the class). And finally, to his surprise, many of his students reported that the use of AI in this manner lowered their anxiety during class. The students appreciated being able to participate in this way. He believes this is because when students see their question listed among the most popular themes, they realize they are not the only ones struggling.

For more details on how Louis did this, see a link in our <u>companion site</u> to Faculty Voices at the Harvard GenAI Library for Teaching and Learning.

3. Summarizing Student Takeaways at the End of Class

Dan has perennially struggled with implementing a technique that he knows is good for his students' learning but that involves too much friction in his teaching. The technique is very simple: asking students at the end of every class a question related to what they learned during that class. This is often referred to as the one-minute paper, and it was also discussed in Chapter 4.

A one-minute paper is a quick and easy way to check in with your students. In just one minute, they jot down their thoughts on a specific question about the class session. It is a great way to see what they are understanding and where they might need more help. It helps you spot any confusion early, tailor your next classes, and keep students engaged. Pedagogically, it encourages students to practice metacognition (i.e., be aware and understand their own thought processes, enabling them to regulate their learning) which is associated with better learning outcomes.

One of the frictions that Dan faced was figuring out what to do with his students' answers to the one-minute paper Initially, he would ask students to write their takeaways on a piece of paper. He would then come back to the office and try to spend enough time sifting through these pieces of paper to extract themes he could share with his students in the next class. He then moved to an electronic system, which made things a bit better logistically but still very time-consuming.

To address this challenge, he got together with Gonzalo Jara, a former student of Dan at the Harvard Kennedy School and now a data scientist. Gonzalo developed a ChatGPT-based platform similar to the one Louis developed (described above) which allowed Dan to implement the following process:

1. Show the students a QR code that they can scan with their smartphone and that leads them to a 2-question survey (their name and a question in the style of "what did you learn today that you want to remember a month from now?").
2. Students spend a minute or two answering the survey.
3. Once enough students have submitted their answers, he clicks a button and within 20 seconds, students see on the screen the 3-5 key themes with the names of the students who contributed to these themes in parentheses (see Figure 6.2 below for an example).
4. Once the answers are displayed, Dan can do a number of things. For example, he could go more deeply into one of the themes, correct any misconception, address another key theme he had hoped students would gain from the class that was not listed, call on a low-participating student to expand on the theme they contributed to, etc. Figure 6.2 provides an example of the type of output produced by the tool Dan and Gonzalo built.

Figure 6.2 - Illustration of Summary of Themes for Key Takeaways (Dan Levy and Gonzalo Jara)

Main Ideas

1. Decision quality and outcome are separate; good decisions can have bad outcomes and vice versa (Mei Chen; Mohammed Hassan; Maria Petrov; Ahmed Ali; Samantha Scott)

2. Decision-making involves considering probabilities and various outcomes, especially in the context of uncertainty and risk management (Luca Rossi; Elena Petrova; Amir Khan; Sofia Lopez; Raj Patel)

3. The decision-making process should be thorough, involving weighing errors of omission against commission and considering heterogeneity and political implications (Svetlana Ivanov; Sophia Kim; Natalia Castro; Valentina Díaz; Gabriel Ramirez)

Source: Output from AI tool developed by Gonzalo Jara and Dan Levy. Themes are based on real answers but the names of the students are fake (to protect the confidentiality of real students).

4. How You Could Summarize Student Input in Class

Louis and Dan developed their own platforms to summarize student input to make the experience a little smoother. However, you can achieve the same output using ChatGPT and the help of someone in your classroom. One possible workflow would be something along the lines of:

- You develop a one-question survey with whatever you want to ask students to submit (e.g., a question, a key takeaway, a point of confusion, etc.). Ideally, use a tool like Google Forms which enables you to quickly retrieve the students' answers.
- You ask students to respond to this question in class on their smartphones or laptops/tablets. Ideally, give them a QR code or a simple link to access.

- Once enough students have answered the survey, you (or ideally someone you designated like a teaching assistant or a student in the class) retrieve the answers to the survey. In Google Forms this is trivial to do as the answers are updated live in a Google spreadsheet.

- You (or your assistant) go to ChatGPT (or a similar AI tool) and paste the student answers along with a prompt you previously developed (along the lines of "summarize the 3 key themes from my students' answers to...") and display the answers from ChatGPT to the students in the class. Then you can take the kinds of actions described above to engage your students and further their learning. This process could be made simpler by building a customized chatbot like the ones we will explore in Chapter 10.

The process described above might not be as elegant as building a platform yourself, but it takes less time, and it works!

If you don't like the idea of developing your own solution or using a teaching assistant to do it live for you in the classroom, keep in mind that companies that produce polling apps (e.g., PollEverywhere, Kahoot, Mentimeter, etc.) will likely develop this capability in their apps soon (hopefully by the time you read this!). Summarizing qualitative answers with AI seems a lot better than through word clouds, their current solution.

Independently of the tool you find most feasible for your needs, we suggest you think of summarizing student input during class as a teaching approach you can now use in your classroom. The two examples above illustrate how you could summarize student questions or student takeaways, but the approach can be used to summarize anything you want. If you think knowing what your students are thinking on a particular issue would help you conduct your class better, this AI-driven summarization approach could be very helpful to you.

5. Other Use Cases of Instructor Use of ChatGPT in the Classroom

The above examples have focused on AI's ability to quickly summarize text during your class (in this case, questions or takeaways from students). We suspect that, by the time you read this, there will be many other applications of ChatGPT in the classroom. In this section, we speculate about possible present and future uses from the instructor or students in the classroom. As a reminder, if you are interested in examples in which *students* are *currently* using AI tools in the classroom, we suggest looking into Part III (especially Chapters 9 and 10).

Here are some other ideas of how you may (now or soon) use ChatGPT in the classroom:

Simulated debate

You may be interested in inviting a guest to your class to represent a different viewpoint than yours, but this may be difficult or costly. Instead, you could ask ChatGPT to play the role of the guest, have a short debate with it, and then debrief with your students. For example, if you are teaching neoclassical macroeconomics, you could ask ChatGPT to represent the Keynesian perspective, and simulate a debate about what to do to reduce the level of unemployment in the country. With the current audio capabilities of ChatGPT, you could have this debate verbally in the classroom (instead of typing back and forth) with your computer or even smartphone connected with your classroom audio system. And with the emerging video capabilities, you could soon be having a debate live with an AI-powered persona on video (though this might take more work on your part). We explore how individual students use ChatGPT in a similar way to deepen their learning in Chapter 8.

Below is a sample prompt.

You are an expert in Keynesian economics. I am an expert in neoclassical economics. We are having a debate in front of a class about how to reduce the level of unemployment in the country. I will present arguments from a neoclassical perspective, emphasizing market efficiency, flexible wages, and minimal government intervention. Your task is to counter my arguments with Keynesian principles, focusing on the importance of aggregate demand, government intervention, and the role of fiscal and monetary policy in addressing unemployment. Let's keep our debate respectful, insightful, and rigorous.

Debate Format:

1. **Opening Statements:** Each side presents their primary viewpoint on reducing unemployment.
2. **Key Points of Contention:**
   o The role of government intervention in the economy.
   o The effectiveness of fiscal policy in reducing unemployment.
   o The impact of wage flexibility on employment levels.
   o The importance of aggregate demand in driving economic activity.
3. **Rebuttals:** Each side responds to the key arguments made by the other side.
4. **Closing Statements:** Summarize the main points and conclude with final thoughts on the best approach to reducing unemployment.

Interactive Case Studies

When you teach a classic case study, the document that students read before class typically establishes a set of facts. Then the classroom discussion refers to these facts as students debate different courses of action. One challenge is that the case stops at a point in time, and it is hard to simulate how each action that you can take might lead to different consequences or reactions from other actors. To address this problem, ChatGPT could generate dynamic case studies on the spot, incorporating new details or facts based on student inputs or changing scenarios.

For example, in a public health class, the instructor could ask ChatGPT to create a case study involving a public health crisis. As students suggest different policy responses, ChatGPT updates the scenario's outcomes in real-time, facilitating a rich discussion. You could do this with a formal classic simulation, but ChatGPT could enrich this simulation by having different actors respond differently (e.g., the governor of your state refuses to enforce a quarantine, whereas the governor of a neighboring state wants to close schools but not bars). See Chapter 9 for a similar simulation designed by Matthew Wemyss.

Enhanced Accessibility

ChatGPT may soon provide real-time support for students with disabilities, offering assistance with note-taking, sign language translation, and other accessibility needs.

For example, in a history class, a hearing-impaired student could use ChatGPT to receive real-time sign language interpretation of the lecture, ensuring they can fully participate in the class.

Live Teaching Assistant

Imagine you could have a second pair of eyes and ears in the classroom that could help you process everything that's happening and assist you by alerting you if something needs your attention. For example, if a student looks confused during a class, the AI tool could notify you to check in with them or provide additional clarification (granted the appropriate permissions, of course). It could also track student participation, ensuring that quieter students get a chance to contribute or alerting you if you seem to be over-calling students on the right side of the room.

This will likely be possible soon. ChatGPT and other AI tools are becoming multimodal, which means they can understand and generate multiple types of data or modes of input and output, such as text, images, audio, and video, rather than being limited to just one type. So an AI-powered tool could be observing the class and assisting you in some of the ways indicated above and many more.

Group Work Assistant

It is common for educators to ask students to discuss something in small groups during class. One of the biggest challenges in this situation is leveraging the learning from those discussions in a productive class-wide debrief. To address this challenge, some instructors ask the groups to document their work. If they do so in a place like a Google Doc or Google slide deck, the instructor can check the work in real-time and prepare for the debrief while students are still working. If you are interested in learning how to use this technique and the benefits it might have, please see Chapter 6 of the book "Teaching Effectively with Zoom" (this chapter is linked in our companion site).

While the above technique works well, AI tools could help even more. Imagine that each group has an AI tool that is listening to the conversation (again with appropriate permissions). Within 30 seconds of the group work being over, a central AI tool could compile the transcript of all group conversations and generate a series of questions for the instructor to ask and which groups to ask these questions from. For example, imagine being able to do something like the following: "Group 3, you argued for position X. Please explain why." followed by "Group 5, you seem to have a contradictory perspective on this than Group 3 because you argued Y. Please expand or reconcile" and so on. This may soon be possible without students having to document their group work (unless you want them to).

💡 Key takeaways

- As you consider using ChatGPT in the classroom, think carefully about the comparative advantages of the classroom environment and how ChatGPT can support these.

- You can use ChatGPT to collect and summarize student questions or key takeaways during class, enhancing participation and providing immediate feedback to address student confusion.

- This chapter presented just initial experimentation of the use of AI inside classrooms and some speculation about future uses. A lot more is coming, so keep an eye on this use case, and experiment yourself!

Chapter #7 - Designing and Grading Assessments

In this chapter, we review how ChatGPT can help you assess your students. The chapter assumes you will keep the same learning goals and types of assignments you had in the past. However, as we will argue in Part III, we believe that the advent of ChatGPT and similar tools should force us to reexamine our learning goals and the way we assess our students. If you want to understand how you may rethink your assessment practices, you can jump to the Introduction to Part III of this book. For examples of how instructors have designed assignments that nudge their students to use ChatGPT for learning, please see Chapter 9.

Below are some of the ways ChatGPT can help you with your current assessment practices.

Table 7.1 - Ways in Which ChatGPT Can Help in Assessing Your Students

Step	How ChatGPT can help you
1 - Design or update an assignment or exam	* Give you feedback on your current assignment * Tweak your current assignment/exam * Brainstorm questions for a new assignment
2 - Design a grading rubric	* Brainstorm learning goals for your assignment * Create a rubric given your learning goals
3 - Give feedback to students or grade	* Assist in providing feedback to students on their work * Assist in the process of grading

1. Design or Update an Assignment or Exam

Designing assignments and exams can be very challenging for instructors. It requires thinking deeply about what we want our students to learn and how we can allow them to demonstrate the extent to which they have done so. While ChatGPT cannot do the thinking for us, it can assist in various ways, including giving feedback on an existing assignment or exam, and brainstorming questions/activities for a new one.

We illustrate these uses with three examples focused on creating relatively long assignment questions or exercises. As a reminder, Chapter 5 also included examples of creating multiple-choice or short essay questions in the context of a pre-class exercise.

Our first example comes from Kimberly D. Acquaviva, a professor at the University of Virginia's School of Nursing, who created a clinical reasoning exercise for a nursing course. She recommends faculty members use a prompting formula that includes the exercise's purpose, learning objectives, context, and any specific instructions or requirements. Below is a sample prompt.

Kimberly D. Acquaviva (University of Virginia)

Generate a clinical reasoning exercise for a nursing course focused on the management of patients with chronic obstructive pulmonary disease (COPD). The exercise should address the following **learning objectives:**

- Identify signs and symptoms of COPD exacerbation.
- Understand the role of nursing interventions in managing a COPD exacerbation.
- Apply clinical reasoning skills to prioritize patient care.

Target audience: Advanced nursing students

Patient scenario: 70-year-old male with a history of COPD presenting with increased shortness of breath, cough, and sputum production

Clinical question(s): Based on the patient scenario, what nursing interventions should be prioritized, and why?

Instructions: Provide a rationale for your prioritized nursing interventions in a short essay format (250-300 words).

Evaluation: Assessment of student responses based on accuracy, completeness, and demonstration of clinical reasoning skills

Feedback and reflection: Guidance on how students can use feedback to improve their clinical reasoning skills in future scenarios

Notice that Kimberly used ChatGPT to complement, not substitute her thinking, as the prompt includes what she wanted to assess (accuracy, completeness, and demonstration of clinical reasoning skills), the question she wanted to ask ("What nursing interventions should be prioritized, and why?), and the type of exercise she wanted (a short essay).

We inserted her prompt in ChatGPT and obtained the answer below with the proposed assessment. We are not medical experts to judge the quality of this clinical reasoning exercise or

ChatGPT's proposed sample student response (at the bottom). However, someone with Kimberly's expertise can judge whether this is a good starting point for further prompting or editing. If it is, Kimberly probably saved a non-trivial amount of time designing this exercise.

 ChatGPT

Clinical Reasoning Exercise: Management of COPD Exacerbation

Patient Scenario:
Mr. John Smith, a 70-year-old male with a history of Chronic Obstructive Pulmonary Disease (COPD), presents to the emergency department with increased shortness of breath, a persistent cough, and increased sputum production over the past two days. He reports that his symptoms have progressively worsened despite using his prescribed bronchodilators and corticosteroids. Upon examination, Mr. Smith appears anxious, is using accessory muscles to breathe, and has an oxygen saturation of 88% on room air.

Learning Objectives:
1. Identify signs and symptoms of COPD exacerbation.
2. Understand the role of nursing interventions in managing a COPD exacerbation.
3. Apply clinical reasoning skills to prioritize patient care.

Clinical Questions:
Based on the patient scenario, what nursing interventions should be prioritized, and why?

Instructions:
In a short essay (250-300 words), prioritize the nursing interventions for Mr. Smith and provide a rationale for each intervention. Consider the urgency of the patient's condition, the goals of care, and the potential outcomes of the interventions.

Evaluation Criteria:
    Accuracy: Correct identification of signs and symptoms of COPD exacerbation and appropriate nursing interventions.

Completeness: Inclusion of all critical nursing interventions and comprehensive rationale for each.

Clinical Reasoning: Logical prioritization of interventions based on patient assessment and clinical guidelines.

Feedback and Reflection:

Guidance: Students will receive individualized feedback highlighting strengths and areas for improvement in their clinical reasoning process.

Reflection: Encourage students to reflect on the feedback and consider how they can apply it to future clinical scenarios to enhance their decision-making and patient care skills.

Sample Student Response:
Prioritized Nursing Interventions for COPD Exacerbation

In managing Mr. John Smith's COPD exacerbation, the following nursing interventions should be prioritized:

1. **Administer Supplemental Oxygen**: Given Mr. Smith's oxygen saturation of 88%, immediate administration of supplemental oxygen is crucial to improve his oxygenation. The goal is to maintain oxygen saturation levels between 88-92% to prevent hypoxia without causing hypercapnia.

2. **Positioning**: Position Mr. Smith in a high Fowler's position to facilitate lung expansion and improve ventilation. This position helps reduce the work of breathing and promotes better oxygenation.

3. **Administer Bronchodilators and Corticosteroids**: Rapid administration of short-acting bronchodilators (e.g., albuterol) via nebulizer or inhaler is essential to relieve bronchospasm. Additionally, systemic corticosteroids (e.g., prednisone) should be administered to reduce inflammation and improve airflow.

4. **Monitor Vital Signs and Respiratory Status**: Continuous monitoring of Mr. Smith's vital signs, oxygen saturation, and respiratory status is vital to assess the effectiveness of the interventions and detect any deterioration early.

5. **Hydration and Mucolytics**: Ensuring adequate hydration can help thin secretions,

making it easier for Mr. Smith to expectorate sputum. Mucolytics may also be considered to further aid in mucus clearance.

6. **Patient Education and Support**: Educate Mr. Smith about the importance of adhering to his medication regimen, recognizing early signs of exacerbation, and when to seek medical help. Providing emotional support can help alleviate his anxiety, which may contribute to his respiratory distress.

In conclusion, these prioritized interventions address the immediate needs of improving oxygenation, reducing bronchospasm, and ensuring effective monitoring and patient education, all of which are critical in managing a COPD exacerbation.

Our second example is based on Dan's experience teaching a statistics course at the Harvard Kennedy School. In 2020, he created a question to assess Bayes rule, an important formula and framework in many introductory statistics courses. The question used COVID testing as an opportunity to apply in a real-world context the techniques they were learning in class. By the fall of 2023, the question felt dated. Dan used ChatGPT to create an updated question.

 Dan Levy (Harvard Kennedy School)

Context: I teach statistics for master's students at a top public policy school that attracts students from all over the world. One of the key goals of my course is that students conceptually understand the statistical methods we study well enough that they are able to apply them to a wide range of public policy issues. One of the classes is devoted to Bayes Rule, and the application we study in class is whether women in their 40s should have regular mammograms. Three years ago, I designed an assignment question in which they had to apply Bayes Rule and more broadly the notion of Bayesian updating to understand how to interpret the results from COVID tests.

Task: The example feels a bit dated and I am hoping I could replace this example with one that feels more current and relevant. Please help me design such an assignment.

ChatGPT

Assignment: Bayesian Updating in Real-Life Public Policy Scenarios

Context:

In the coastal city of Portsville, a recent survey suggested an increase in water contamination levels. Authorities suspect the presence of a specific pollutant, called ChemX. This pollutant, when consumed, has been known to lead to a unique set of symptoms in humans. To assess the risk, the health department of Portsville begins to test residents for the presence of ChemX in their system. A standard test has been developed that detects ChemX's presence, but like all tests, it's not perfect.

Data provided:
- Base rate (prior probability): Only 5% of the general population in similar cities without known contamination has traces of ChemX in their system.
- Test accuracy: The test has a 95% true positive rate (if someone has ChemX, there's a 95% chance the test will be positive) and a 90% true negative rate (if someone doesn't have ChemX, there's a 90% chance the test will be negative).

Questions:

1. **Calculations and Base Rate:**

- A random resident, Alice, has been tested and her result came back positive for ChemX. Using Bayes Rule, what is the probability that Alice actually has ChemX in her system?

Hint: This will require you to use the base rate as your prior, and update it using the likelihood of the test result.

2. **Interpretation**:
- Another resident, Bob, received a negative test result. How confident should Bob be that he does not have ChemX in his system?
- Discuss the importance of understanding the true negative rate in making this assessment.

3. **Conceptual Understanding**:
- The health department considers conducting a campaign to educate the residents on the ChemX issue, the prevalence rate, and the test's accuracy. Why is it essential for the general public to understand the base rate and the test's accuracy? And how might public reactions differ if they only knew about the test's accuracy and not the base rate?

4. **Real-life Policy Implication**:
- Imagine you're advising the Mayor of Portsville. Considering the base rate of ChemX presence in similar cities, would you recommend issuing a city-wide advisory based on a 10% positive rate from the first 1,000 tests? Why or why not?

5. **Bayesian Updating Perspective**:
- The health department conducts a second test on individuals who tested positive in the first round. This second test has a 98% true positive rate and a 93% true negative rate. If Alice, from Question 1, tests positive again in the second test, how should she update her probability of having ChemX in her system?

Hint: Use the result from Question 1 as the new prior for this Bayesian update.

This assignment should allow your students to flex their conceptual, computational, and interpretative muscles on Bayes Rule, base rates, and the application of these principles in policy decisions.

Dan was very impressed with ChatGPT's answer. Many of the questions included in the proposed assessment were as effective at testing students' understanding of Bayes Rule as his original assessment. However, he wanted to make some changes to tailor the questions further to the context of his students and his learning goals (see below).

 Dan Levy (Harvard Kennedy School)

Ok. This looks good in general. But a few changes please: (1) Students come from all over the world, so please make the example a bit more international. (2) Please use a policy area other than health since I already use a health example during class, (3) To the extent possible, rather than referring to generic names like ChemX, use specific real names for things, and (4) In the "Data Provided" section, please present information as students will hear about it in the real world (i.e., not using language such as prior probability and base rate)

 ChatGPT

Assignment: Bayesian Updating in Real-Life Public Policy Scenarios

Context:
The World Cultural Heritage Commission has recently raised concerns about potential forgeries in global art markets. Notably, there's been a sudden emergence of paintings believed to be by the famous Italian Renaissance artist, Leonardo da Vinci. While some of these paintings could be genuine undiscovered masterpieces, others might be sophisticated forgeries. The International Art Council (IAC) has developed a specialized

test, the "Renaissance Authenticity Test" (RAT), to assess the legitimacy of these paintings. The world looks to the results of this test to decide on the authenticity of the art pieces.

Data provided:
- Out of every 1000 paintings from the Renaissance era found in the last century, 10 were authentic Leonardo da Vinci pieces.
- RAT accuracy: When presented with a genuine da Vinci, the test confirms its authenticity 95% of the time. On the other hand, when presented with a forgery, the test correctly identifies it as a forgery 90% of the time.

Questions:
Calculations and Initial Information: A newly discovered painting, "La Donna Del Lago", was tested using the RAT and its result came back confirming it's an authentic da Vinci. Using the provided data, what is the probability that "La Donna Del Lago" is indeed a genuine masterpiece by Leonardo?

Hint: You'll need to consider the number of genuine da Vinci paintings out of every 1000 Renaissance era discoveries.

[.... Selected questions omitted for brevity...]

This revamped assignment should offer a more international and non-health-related context, and the detailed real-world information should engage your students effectively.

The example that ChatGPT came up with startled Dan. After 20 years of teaching statistics, he would have never come up with an example of potential forgeries in global art markets. While he appreciated ChatGPT's versatility and greater breadth than his own, he wanted an example that felt more policy-relevant for his students. Hence, he adopted a different approach that tends to be more effective when you are still in the brainstorming phase. Rather than having ChatGPT come up with a single full example, he asked ChatGPT to generate several ideas of examples first, and then only once he settled on one, he asked ChatGPT to develop it into a full assignment question.

 Dan Levy (Harvard Kennedy School)

Great. I like the structure of the prompt and the questions you included in it. Can you please come up with 5 more policy-related examples where we could use this structure of questions? For now, just describe each example in 2-3 sentences. I will then select one of the examples you came up with and ask you to write the prompt and the questions for that example.

 ChatGPT

Certainly! Here are five policy-related examples:

1. **Archeological Authenticity**:
   - International archaeologists have uncovered numerous clay tablets in the Middle East purported to be ancient Sumerian writings. However, there's concern that modern forgeries have infiltrated these discoveries. A special analysis method has been developed to determine the age and authenticity of these tablets.

2. **Fishery Sustainability**:
   - Overfishing is a concern in international waters. Various fish species have been marked 'at risk', and there's a new sonar technology that claims to determine the species and age of fish schools to ensure fishing boats are adhering to international standards. The accuracy and implications of this technology are under scrutiny.

3. **Deforestation Monitoring**:
   - With rising concerns about deforestation in the Amazon, satellite imaging is being used to identify legal versus illegal logging activities. However,

distinguishing natural clearings from man-made ones isn't always accurate. The effectiveness and trustworthiness of these satellite images are pivotal for international environmental policies.

4. **Antique Firearms**:
   - As antique firearms fetch high prices in global auctions, a surge in potential "historic" firearms from battles like Waterloo or Gettysburg has emerged. A new metallurgical test claims to date these firearms accurately. The reliability of this test impacts both cultural heritage considerations and market values.

5. **Cultural Heritage Site Authenticity**:
   - Several sites are vying for UNESCO World Heritage status. Some claim to be ancient sacred grounds or significant cultural crossroads from millennia ago. Geographical and historical tests are employed to verify these claims, and their accuracy plays a role in international recognition and tourism policies.

You can choose one, and I'll be happy to provide a detailed assignment based on it.

Dan liked the deforestation monitoring example and asked ChatGPT to draft a full assignment question with this case as the backdrop. We recommend that you follow this process (brainstorm first, then flesh out a specific idea) when you are updating or creating an assignment from scratch with ChatGPT. Even though it took some back and forth, Dan estimates that ChatGPT saved him a couple of hours of work, and he ended up with a much richer set of examples to choose from than if he had tried doing this on his own.

If ChatGPT can create assignment questions, it's natural to think it could also answer them for students. We address the challenge of ChatGPT potentially hindering learning in Part III of the book. For now, we suggest that students are more likely to use ChatGPT to complete assignments if they lack motivation or find the assignment irrelevant.

Some educators argue that AI can help create more engaging and relevant assignments for students. For example, Danny Liu, professor at the University of Sydney in Australia, wondered,

"Instead of trying to 'AI-proof' assessments, can AI help us to uplift our approach to designing and delivering assessments and feedback that students care about?"

Below is an example from Danny. His prompt below is designed to generate creative assessment ideas for a set of learning outcomes. The prompt also asks for assessments that encourage gradual work. Danny argues that this is consistent with the notion of assessing the process, and not the product. With the advent of generative AI, focusing our assessment practices on process seems wise.

 Danny Liu (University of Sydney, Australia)

Suggest 5 creative assessment ideas that can assess the following learning outcomes in a university-level media and communications course. Make the assessment meaningful for students, encourage gradual work towards the final product, and use non-essay formats.

LO1. demonstrate a broad yet nuanced understanding of introductory theories in media studies

LO2. demonstrate how to apply these theories to 'reading' media texts

LO3. critically evaluate the usefulness of theories in addressing questions about media production and reception

LO4. demonstrate skills in academic writing and research

LO5. demonstrate personal and intellectual autonomy through assessment and class work.

 ChatGPT

Here are five creative assessment ideas for a university-level media and communications course that align with the specified learning outcomes:

1. Theoretical Frameworks Podcast Series
- **Description:** Students create a series of short podcast episodes (3-5 minutes each) explaining key media theories, their origins, and their applications. Each episode should include a brief discussion of a specific media text (e.g., a film, TV show, or news article) analyzed through the lens of the theory discussed.
- **Learning Outcomes Addressed:** LO1, LO2, LO5
- **Structure:**
  - **Week 1:** Choose theories and media texts.
  - **Week 2:** Research and script the first episode.
  - **Week 3:** Record and edit the first episode.
  - **Subsequent Weeks:** Repeat for remaining episodes, with feedback sessions in between.

2. Media Theory Infographic Portfolio
- **Description:** Students design a series of infographics that visually represent different media theories, their key concepts, and how they apply to analyzing media texts. Each infographic should include a case study of a media text and a critical evaluation of the theory's usefulness.
- **Learning Outcomes Addressed:** LO1, LO2, LO3, LO5
- **Structure:**
  - **Week 1:** Select theories and media texts.
  - **Week 2:** Draft initial infographic layouts.
  - **Week 3:** Create the first infographic using digital tools.
  - **Subsequent Weeks:** Develop additional infographics, with peer reviews and iterative improvements.

[...Selected content omitted for brevity...]

2. Design a Grading Rubric

A grading rubric is a tool that clearly outlines the criteria for assessing student work, specifying different levels of performance for each criterion. They can be helpful because they provide transparency for students about their assessment, set clear expectations, and ensure consistency in grading. Designing a good rubric, however, can be time-consuming and challenging. ChatGPT can help you brainstorm your assignment learning goals, create a rubric given your learning goals, or give you feedback on an existing rubric.

Below is a sample prompt to create a rubric with ChatGPT, suggested on a website called AI for Education (linked in our companion site). It was slightly modified for brevity's sake. It is useful that it asks ChatGPT to generate the rubric in a table format, and that it asks as well for descriptions of each criterion on a 4-point scale.

 AI for Education *[Sample prompt for creating rubric]*

You are an expert teacher and curriculum writer, skilled in creating assessments and evaluating student work. Your task is to create a rubric for my 5th grade science class studying engineering. My students are doing a STEM experiment, the "egg drop challenge," in which they use specific materials to design and build a container to protect the egg from cracking as it is dropped from increasingly higher heights. Format the rubric as a chart and include a 4-point scale.

 ChatGPT

Sure, here is a detailed rubric for the 5th grade STEM experiment "Egg Drop Challenge," formatted as a chart and using a 4-point scale:

[Rubric pasted in table below]

Descriptions of Criteria:

- **Design and Creativity**: Evaluates the originality and innovativeness of the design, as well as how effectively the materials were utilized.
- **Structural Integrity**: Assesses how sturdy and well-constructed the structure is to protect the egg.
- **Functionality**: Measures the success of the design in protecting the egg through various heights of drops.
- **Engineering Process**: Looks at the students' understanding and application of engineering principles such as planning, designing, testing, and refining.
- **Teamwork and Collaboration**: Evaluates the effectiveness of teamwork, including how well students worked together and shared responsibilities.
- **Presentation and Explanation**: Assesses the clarity and organization of the presentation, as well as how well the students explained their design, process, and results.

This rubric provides a comprehensive assessment of the students' performance in the "Egg Drop Challenge," covering all aspects from creativity and design to teamwork and presentation.

Table 7.2 - Grading Rubric for STEM Assignment on "Egg Drop Challenge"

Criteria	4 - Excellent	3 - Good	2 - Satisfactory	1 - Needs Improvement
Design and Creativity	Highly innovative design, exceptional creativity, and effective use of materials	Creative design with good use of materials	Design shows some creativity, moderate use of materials	Limited creativity in design, basic use of materials
Structural	Structure is very	Structure is sturdy	Structure has	Structure is weak

Integrity	sturdy and egg is well protected	with minor weaknesses	some weaknesses but still protects the egg	and barely protects the egg
Functionality	Egg survives all drops without any damage	Egg survives most drops with minor damage	Egg survives some drops with moderate damage	Egg does not survive drops or is severely damaged
Engineering Process	Demonstrates thorough understanding and application of engineering principles	Good understanding and application of engineering principles	Basic understanding and application of engineering principles	Limited understanding and application of engineering principles
Teamwork and Collaboration	Excellent collaboration, everyone contributed effectively	Good collaboration, most team members contributed	Satisfactory collaboration, some team members contributed	Limited collaboration, few team members contributed
Presentation and Explanation	Clear, well-organized presentation with thorough explanation	Good presentation with clear explanation	Satisfactory presentation with adequate explanation	Presentation lacks organization or is missing key explanations

Below is a prompt for creating rubrics from Danny Liu, professor at the University of Sydney in Australia. The prompt will have ChatGPT ask you a few questions to find out more about what you need, before generating a draft rubric for you. Danny reported: "The rubric that it generates is by no means something I would immediately use for students, but it gives me things I can think about and improve upon."

Danny Liu (University of Sydney)

Act as an expert higher education academic and writer of assessment rubrics.

I need to write a marking rubric for a second-year university-level assessment where students (pre-service teachers) need to critique the use of technology in high schools. You need to help me generate the rubric rows (which are the criteria), and the columns are the various standards (from high distinction, to distinction, to credit, to pass, to fail).

Ask me questions to help me write a strong rubric. Ask one question at a time, wait for my response, and ask me the next question. After I have answered your questions, produce a draft rubric for me.

If you are interested in using ChatGPT to create grading rubrics, there are plenty of resources on the web to do this, including some Custom GPTs that can assist you in the process! See Chapter 10 for more on Custom GPTs.

3. Give Feedback to Students or Grade Student Work

Providing feedback to our students is crucial for their learning and development, as it helps them understand their strengths and areas for improvement. However, crafting detailed and personalized feedback can be extremely time-consuming. ChatGPT can assist in this process by generating initial feedback based on student submissions, helping you quickly draft comments that you can then refine and personalize.

Nonetheless, there are several drawbacks associated with using ChatGPT to provide feedback to your students. First, feedback generated by ChatGPT might lack the nuanced understanding and personal touch that only you, as the instructor, can provide. At the end of the day, you are the expert, and you know best what feedback to give and how to provide it. Second, you want to ensure that the feedback aligns with your teaching objectives and assessment criteria, which requires careful oversight of ChatGPT or any AI tool that you use. Third and perhaps most importantly, relying heavily on AI-generated feedback could lead to a sense of disconnection between you and your students. Students might perceive the feedback as less genuine or less tailored to their individual needs. Some are motivated by the fact that their instructor will be reading and reviewing their work, and may feel discouraged if AI does this job instead. In fact, our sense from our conversations with many students is that they generally don't like the idea of instructors using AI to provide feedback. Some even see this practice as an abdication of the instructor's responsibilities. Others very reasonably argue that they could get feedback from an AI tool themselves, and question the value added of the instructor doing it. Finally, some educators have pushed back against the idea that feedback and grading are activities to be automated, so that instructors can have more time for their students. They have argued that grading is integral to their relationships with their students, and that automating the feedback and grading degrades those relationships.[8]

If you still would like to obtain assistance providing feedback to your students or grading, we subscribe to the advice given by Bruce Ellis (2023)[9], slightly modified for our purposes:

1. **ChatGPT does not know your students**. It lacks the context and understanding of their needs, strengths, and circumstances.
2. **ChatGPT isn't perfect**. It's true regardless of the task that you are asking it to help with. Using ChatGPT to help grade is no different. You will need to review what it generates, edit, revise, and tweak.
3. **You can't rely solely on ChatGPT for grading**. Can it help? Yes. But you should not take its assessments as final.
4. **You are ultimately responsible**. You are responsible for the accuracy of the grading and the feedback. You will need to be able to defend the grades and feedback you give students.

We finish this chapter with two examples from educators who have used ChatGPT to give feedback to their students. Bruce Ellis, Director of Professional Development with Texas

Computer Education Association (TCEA), used ChatGPT to assist with grading college essays. He uploaded the grading rubric, followed by each student's essay, and then asked ChatGPT to suggest feedback based on the rubric. As a reminder, the regular version of ChatGPT uses your prompts for training purposes, so we recommend having access to a ChatGPT version (e.g. ChatGPT Enterprise) or some other AI tool with the right privacy protection if you need to use sensitive data.

 Bruce Ellis (Texas Computer Education Association)

Act as an expert educator who is able to read information, analyze text, and give supportive feedback based on a rubric that I will give you.

Students were given the following assignment:

[paste student assignment here]

When you are ready, I will give you the rubric. I will then begin to give you the student work to evaluate and you will provide specific, constructive, and meaningful feedback in a first-person, supportive voice. If the score given is less than **[total points for the rubric]** points, provide a brief paragraph on specific steps the student can do to improve the work and earn full credit based on the rubric.

Do you understand?

Another example comes from Jason Tangen, professor at the University of Queensland in Australia. Jason created a prompt that suggests full-fleshed ideas for improvement based on short pieces of feedback that he types. Jason reports: "For example, you may have made a brief note on a student's essay stating 'Argument lacks clarity and specificity.' When provided to ChatGPT, along with the essay prompt and rubric, it might expand this feedback to say, 'The main argument of your essay could be further enhanced by providing more specific examples and explanations. Remember that a clear and specific argument helps the reader understand your point of view better and strengthens your overall position.'" If you are interested in pursuing this further, the companion site has a link to Jason's description of the system and his full prompt. It also has a link to an interesting example of using ChatGPT to provide batch feedback on student writing.

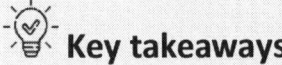 **Key takeaways**

- **ChatGPT can help you design or improve an assignment or exam.** It can give you feedback on an existing assignment, help you tweak or change the context of an assignment, or assist you in creating new assignments.
- **ChatGPT can help you design grading rubrics.** It can help you brainstorm learning goals for your assignments, assist you in creating rubrics based on those goals, and provide you feedback on your rubrics.
- **ChatGPT can assist you in providing feedback to your students and in helping grade their work.** While using ChatGPT in this way can help you save time and perhaps provide more extensive feedback, consider the drawbacks involved before deciding whether to follow this path.

Part III - Ways Your Students Can Use ChatGPT

In the previous part of this book, you hopefully gained a sense of how powerful ChatGPT can be to assist *you* in teaching more effectively and in helping *you* be more productive at work. Similarly, ChatGPT can also help *your students* learn more effectively, be more productive, and develop skills that will be valuable in their future jobs.

A lot of the discussion about students using ChatGPT (or generative AI more generally) has centered around cheating. The notion is that students can use ChatGPT to do the work for them, rather than with them. There is no doubt that this will hinder learning rather than enhance it.

This part of the book focuses on how our students can use ChatGPT to enhance their learning (Chapter 8), and how we as educators can nudge our students to do so (Chapters 9 and 10). But before we go there, let's walk together through an example that illustrates some of the central tensions that student use of ChatGPT brings.

1. A Grounding Example

Dan teaches a course on analytic methods at the Harvard Kennedy School. He has a class session on a statistical framework called "Signal Detection Theory" which is used to distinguish between signals (i.e., information-bearing patterns) and noise (i.e., random patterns that distract from the signal). For example, signal detection theory is used to improve how well your email can distinguish between legitimate emails (signals) and spam that mimics genuine messages (noise).

Dan wanted his students to learn about signal detection theory before class so that during class he could skip the basics, and his students could delve more deeply into the subject and its applications in the real world. Before ChatGPT came along, his pre-class assignment consisted of a reading followed by a short survey:

Table III.1 - Dan's Assignment about Signal Detection Theory in the Pre-ChatGPT Era

Activity	Estimated Time
Read Chapter 7 of Steven Pinker's "Rationality" book	60 minutes
Answer Survey	15 minutes
Total Estimated Time	75 minutes

Before you go any further, we want you to pause and ask yourself: In what ways could Dan's students use ChatGPT to complete this assignment and be ready for class?

For many of us, the first use that comes to mind is that students could skip the reading and use ChatGPT to answer the two questions for them. This would reduce the amount of time it would take to complete the pre-class assignment from 75 minutes to about 3 minutes. It would also reduce the learning from whatever was achieved before ChatGPT existed to about zero! In fact, Dan tried to assess how good ChatGPT is at this task, and he estimated that ChatGPT's answers were at least as good as the average student in his class.

Before we delve deeper into this type of use and its implications for education, we invite you to think about other possible uses of ChatGPT in this context. Here are some we came up with to complete the first activity (reading the text):

- Students could use ChatGPT to familiarize themselves with the basics of the reading before reading the full text, to get more out of the reading, and perhaps do so more quickly.
- Students could ask ChatGPT to summarize the text for them, and hence reduce the time it would take them to become familiar with the text (at the expense of not getting as deep an understanding of the subject). They could follow up by asking ChatGPT questions about the summary (e.g., Can you explain the second bullet a bit more? Can you give me examples of the third bullet in my area of interest? etc.).
- Students could read the text and ask ChatGPT questions about things they did not understand or want to know more about.
- Students could ask ChatGPT to teach them about the content of the text (i.e., signal detection theory) in a personalized manner that adapts to their background on the topic, their area of interest, and how they like to learn (see sample prompt in Figure III.1 further below).

With regards to the second part of the assignment (answering the survey), the main two items in the survey were the following:

1. Please describe a decision in your policy area of interest that would benefit from the framework of signal detection theory. [2-3 sentences]
2. Please characterize the key tradeoff in the decision you described above using the framework of signal detection theory. [one short paragraph]

Here are some potential uses of ChatGPT other than simply using it to answer the questions:

- Students could answer the questions themselves and ask ChatGPT for feedback or grammar assistance. Students would end up with a better answer and hopefully some learning in the process.
- Students could come up with an area of interest and ask ChatGPT to provide examples within that area and then through a series of questions develop a complete answer.
- Students could ask ChatGPT to answer the survey for one policy area (e.g., education), learn from ChatGPT's answer, and then try to answer the questions for another policy area (e.g., health).

Hopefully, you will notice that not all of these uses of ChatGPT hinder learning to the same extent as skipping the reading or using ChatGPT to answer the questions. In fact, some may actually increase learning. It's not as simple as "ChatGPT is good for learning" or "ChatGPT is bad for learning." ChatGPT can both aid and hinder learning, depending on how it's used. Our brains need time to process and make meaning to learn effectively. As Willingham (2009) aptly puts it, "Memory is the residue of thought."[10] When students use ChatGPT to ask questions, make sense of the material, explore and probe, it can enhance their learning. However, if they use it merely to find answers to assignments, it won't. This leads us to a key idea of Part III of the book:

Key idea - **Different Uses of Generative AI can Lead to Different Degrees of Learning**

We now invite you to enter the prompt below (also, on our <u>companion site</u>, in case you would like to copy and paste) in your ChatGPT chat box. Tweak it, and spend 5 minutes trying to learn about signal detection theory. Even if this subject is of no interest to you, we think this exercise will be useful in helping you understand how ChatGPT could be used to learn any subject. Make sure you have an exchange (akin to a conversation) with ChatGPT after your initial prompt in which you ask questions specific to your background and interests. Jot down your impressions of how it feels to learn about a subject using ChatGPT as a tutor.

Figure III.1 - Sample Prompt to Learn About Signal Detection Theory from ChatGPT

Template:

- I would like to learn about signal detection theory
- I HAVE / DON'T HAVE much background on this topic
- Please use examples in THIS AREA [FILL IN AN AREA OF INTEREST]
- Explain to me as if I were a HIGH SCHOOL/COLLEGE student
- I learn better in the FOLLOWING WAY... [FILL IN THE BLANK]
- Please give me an explanation of 1/2/3/4 paragraphs
- Quiz me THROUGHOUT / AT THE END to make sure I understood

Sample prompt:

"I would like to learn about signal detection theory. I don't have much background on this topic. Please use examples in the health field. Explain to me as if I were a high school student. I learn better through examples. Please give me an

---

We hope this experience led you to realize that ChatGPT allows for personalized learning in a way that would be very hard to achieve through other means. The reading that Dan assigned, as good as it was, did not allow for this kind of learning. Since the text is the same for all students, it cannot be personalized for each of them. The same principle applies to many other ways in which we try to help our students learn (class sessions, homework assignments, etc.). What ChatGPT (and generative AI more generally) allows is akin to having a personal tutor available 24/7, that can answer specific questions that students have, using the level/language most appropriate to each student's background knowledge, and using examples that interest them.

If you think this example is merely theoretical, we want to try to nudge you to think about content or ideas in your courses that would benefit from this kind of personalized learning, and experiment with them. The example below illustrates how you might leverage the power of personalization in your teaching. In one of Dan's courses, he asked students to use ChatGPT before class to learn about four risk management strategies: options, hedge, diversification, and quitting. Dan felt the need to do this because in his classroom there were students ranging all the way from having worked in finance for several years (who would be very familiar with these strategies) to some who had never heard of these terms before. Giving them the same pre-class assignment would lead to the first group getting bored or the second group getting lost. So in the pre-class assignment, students were given a very brief description of these 4 risk management strategies and encouraged to personalize their learning of these strategies. Sean Norick, one of the students in the class, wanted to emphasize intuition in the explanation and had the exchange below with ChatGPT.

Sean Norick (Harvard Kennedy School Student)

You are a world renowned tutor and are tasked with helping me to learn 4 key risk management strategies: options, hedging, diversification and quitting.

Section A - Basic tools for risk management

The following are strategies often used to manage risk in the world of finance:

Options: Pay a small amount of money to have the option to buy (or sell) a stock at a guaranteed future price. If the stock price plummets (or soars), the trader can exercise the option and make a profit. If it doesn't, the trader loses only the small amount he/she/they paid for the option. It has a payout profile with large upside, but small downside.

Hedge: An investment that is made in an 'offsetting position' with the intention of reducing the risk of adverse price movements in an asset.

Diversification: A risk management strategy that mixes a wide variety of investments within a portfolio. A diversified portfolio contains a mix of distinct asset types and investment vehicles in an attempt at limiting exposure to any single asset or risk.

Quitting: Being able to recognize the 'bad' outcome has happened and quickly exiting your position. There is a famous saying in finance: "cut your losses short and let your winners run".

===================================

You have 15 minutes to make me remember the material forever. It can be a back-and-forth convo if you wish. Go.

ChatGPT

Alright! Let's make this engaging and memorable. I'll use a combination of analogies, questions, and interactive elements to help you internalize these concepts.

1. Options: The Concert Ticket

Imagine you're a huge fan of a band, and they might be coming to your city. You're not sure if you'll be able to attend, but you don't want to miss out if you can. So, you pay a small fee to reserve a ticket at a set price. If the concert becomes super popular and ticket prices skyrocket, you can buy your ticket at the price you reserved. If you can't go or decide not to, you only lose the small reservation fee. This is how options work in the stock market.

Question: If the concert becomes unpopular and ticket prices drop, would you use your reservation to buy the ticket at the higher price?

Notice how the instruction to help Sean "remember the material forever" prompts ChatGPT to use a practical example that Sean may relate to. In this case, the example is about paying fees to reserve tickets for a concert. Note that if Sean was not a big music fan, he could have asked ChatGPT to provide a different analogy. Additionally, ChatGPT included a question to assess Sean's understanding of the concepts. Sean proceeded to have a back-and-forth conversation with ChatGPT that covered all four strategies mentioned.

Now let's look at a different example from Ryan Silber, another student in the class, who faced the same assignment, but who was already familiar with the topic.

 Ryan Silber (Harvard Kennedy School Student)

I am a student enrolled in a masters program at a top public policy school.

My professional experience before enrolling in the masters degree was in climate change

> and urban planning.
>
> I would like to learn about risk management, in particular about options, hedging, diversification and quitting as tools to manage risk.
>
> This is for a course I am taking titled "Thinking Analytically in an Uncertain World"
>
> I have a strong background on this topic. Explain to me as if I were a PhD student.

Note that while Sean had emphasized he wanted the instruction to be memorable, Ryan focused on asking ChatGPT to provide a technical answer that would fit better his mastery level. Also note that, even without an explicit request, ChatGPT provided an explanation relevant to Ryan's areas of expertise: climate change and urban planning.

 ChatGPT

Certainly, I'd be happy to provide you with an advanced overview of risk management with a focus on options, hedging, diversification, and quitting strategies. Since you have a strong background in climate change and urban planning, I'll approach this topic with the assumption that you are already familiar with some key concepts in risk management.

1. **Options:**
   - **Financial Options:** Options are financial derivatives that grant the holder the right but not the obligation to buy (call option) or sell (put option) an underlying asset at a predetermined price (strike price) on or before a specified expiration date. They are commonly used to hedge against price fluctuations or to speculate on future price movements.
   - **Real Options:** In the context of climate change and urban planning, real options can be applied. Real options are the application of financial option principles to tangible assets, such as infrastructure projects. For example, when considering the construction of a flood-resistant infrastructure project, a real option might involve the decision to build it now or delay the

construction based on updated climate change data.

2. **Hedging:**
   - **Portfolio Hedging:** In a policy context, hedging can be essential to manage risks associated with climate change policies or urban planning projects. Portfolio hedging involves using financial instruments to offset the potential losses or gains in a specific project or policy due to uncertain factors. For example, if you're implementing a policy to reduce greenhouse gas emissions, you might use carbon futures or options to hedge against the risk of carbon price fluctuations.

3. **Diversification:**
   - **Systematic vs. Unsystematic Risk:** Diversification is a technique used to reduce risk by spreading investments or actions across various assets or projects. In the context of climate change and urban planning, systematic risk refers to factors like global climate trends that affect all projects, while unsystematic risk relates to project-specific factors. Diversification can help mitigate unsystematic risk by having a diverse portfolio of projects or policies.

4. **Quitting Strategies:**
   - **Real Options Valuation:** Knowing when to quit or abandon a project or policy is a critical aspect of risk management. In real options analysis, the concept of "quitting" is often evaluated as an option itself. This is called the "abandonment option." You assess at what point it might be economically rational to abandon a project in response to unfavorable developments or new information. This is particularly relevant in long-term urban planning projects where the environment and circumstances can change significantly over time.

In the context of your "Thinking Analytically in an Uncertain World" course, you can apply these concepts to analyze and make informed decisions in the face of complex, uncertain, and dynamic situations. Consider employing quantitative models, simulations, and decision trees to assess the potential outcomes of different strategies, and factor in external variables such as climate data, policy changes, and market conditions.

As a PhD student, you can delve deeper into these topics by exploring advanced models like the Black-Scholes model for financial options, advanced risk management techniques such as Value at Risk (VaR) and Conditional Value at Risk (CVaR), and real options models that consider stochastic processes and more complex decision-making scenarios. Ultimately, the key to successful risk management is understanding the interplay of these tools and applying them judiciously to your specific policy and planning challenges.

While this explanation may not have resonated with Sean, it was tailored to fit Ryan's context and learning needs. There is little in common in both explanations and yet both Sean and Ryan indicated that ChatGPT was highly effective in helping them learn about risk management. It would be really hard for an educator to tailor their instruction to two students as different as them.

There are multiple other ways to use ChatGPT to personalize the learning process. For example, one student at the Harvard Kennedy School explained how they blended the use of ChatGPT with in-class instruction to enhance their learning process: "Many times, during classes, I ask ChatGPT to explain to me what the professor is talking about. For example, I take a print screen from slides and ask what each parameter and variable means and what they are trying to tell you that is not written (usually it is also not well explained in the book)." You might not approve of this use of ChatGPT in class, but we suspect it would be hard for you to deny that you wish you could have done this in some of the courses you took when you were in college or graduate school! A different student highlighted that getting a general understanding of a concept or theory with ChatGPT before doing an in-depth reading of a textbook was helpful for their learning. Furthermore, students can use ChatGPT to explain complex graphs or technical explanations in simple words at any time as they are going through the course material.

When we asked our students what the experience of learning with generative AI as a tutor feels like, they reported that they can ask the AI tutor any question they want without fear of embarrassment of how basic the questions might be. They also see as an advantage that the tutor is available at any time (say at 1 am if they are doing homework at that time) and that the tutor has "infinite patience."

There is a deeper pedagogical reason why we think that learning with the help of ChatGPT can be so helpful. Because the nature of the engagement with ChatGPT is akin to having a conversation, it invites students to play an active role, to ask questions, to wonder and inquire when they don't understand something, to be curious, and ultimately to have more agency over their learning process. We see this as positive and consistent with the active learning principles behind the science of learning highlighted in Chapter 2.

In sum, when used appropriately, ChatGPT can be a remarkably powerful tool for learning. However, we of course acknowledge that not all students will use it appropriately. Let's now address the legitimate concerns regarding the use of ChatGPT in learning.

2. Concerns About Using ChatGPT for Learning

Given the benefits described above, are we advocating for an education without teachers or teaching assistants? Absolutely not. However, we think AI can be leveraged as a complement to the very human experience of learning in a physical classroom from a devoted educator. Now let's proceed to the very valid concerns of trying to leverage generative AI for learning. We will focus on three concerns that we think are particularly relevant. However, we want to acknowledge upfront that there are several other concerns related to the use of AI, including privacy, intellectual property, existential risks, the potential widening of the digital divide, and many others which fall outside the scope of this book.

As you read these concerns, you might ask "What should I do?" The last two sections (below) provide some policy options and guidance for you to address them.

Concern #1 - Students can use AI for cheating

Most of us would agree that a student using ChatGPT to do an assignment for them (as in the example above) constitutes a form of cheating. In "Cheating Lessons," one of several wonderful books that Jim Lang has written about teaching, he argues that cheating is often a symptom of educational environments that prioritize grades and outcomes over genuine learning. He suggests that when students are placed in high-pressure situations with extrinsic motivators, such as grades or test scores, they are more likely to cheat. Lang's analysis led him to conclude that educators can reduce the incidence of cheating by designing courses that emphasize intrinsic rewards, deeper understanding, and the learning process itself.

Regardless of how much you agree with Lang or how much you think you can apply his lessons to your own educational context, we argue that examining the forces that might lead a student to cheat will reveal that ChatGPT represents the latest vehicle through which cheating can happen, but it is not the driver of cheating. It certainly makes cheating easier but trying to control student use of ChatGPT is not likely to be successful.

When we asked students when they were more likely to use ChatGPT to do things their professor would not approve of, a common theme was that they do it when they don't think the assignment they are asked to complete is meaningful or worth completing. You might argue that students are not in the best position to know what is meaningful or worth doing because they lack the expertise to do so or because they might favor short-term benefits over long-term ones. And this might well be true on many occasions. But whether we like it or not, with the availability of ChatGPT and other AI tools, students have a lot more agency over how much effort to exert on the different assignments we ask them to do.

Concern #2 - AI can mislead students by providing false or biased information

As described in the preface, ChatGPT is known to produce content that sounds true but is not, leading some to worry that its "hallucinations" may lead our students (and ourselves) astray. ChatGPT is trained on a large corpus of text, including most of what's on the internet, so its output reflects many of the biases that are present online. There are plenty of examples you can find online about how ChatGPT or some other AI models produced some answers that were inaccurate or biased.

Given the goals of education, where fostering critical thinking and ensuring the accuracy of information is vital, this concern is particularly significant. We suggest, however, to keep the following points in mind as you think about this issue:

- **Foster critical thinking skills:** Just like us, students will need critical thinking skills to enable them to determine how to verify and evaluate the information they encounter from many sources in the world (internet, social media, a friend, etc.). These skills are equally important when evaluating information that ChatGPT gives them. They will need these skills when they use AI in their jobs, so learning how to do this during their schooling would likely be valuable for them.

- **Think about the counterfactual:** If your students had access to you 24/7, they would likely get more accurate answers than ChatGPT's. But they don't. So take a moment to consider their other sources of learning (course materials, teaching assistants, classmates, the internet, etc.) and assess the extent to which they are more accurate than ChatGPT. We argue that some are likely to be less accurate than ChatGPT for at least some of your course's content. Therefore, it is helpful to consider how your students would have learned in the absence of using ChatGPT, and the extent to which ChatGPT might be more inaccurate than the other sources they might use.

- **Assess the areas in which ChatGPT is most likely to mislead students:** ChatGPT will likely be more reliable for content for which there is a large corpus of widely accepted information available in the world (e.g., introductory ideas in history, economics, statistics, etc.), than for content in niche fields, or related to widely controversial or complex issues. Depending on what you teach, the concern of inaccurate information from ChatGPT might be more or less salient.

- **Assess costs of inaccurate information:** Students getting inaccurate or biased information is more consequential in some domains than others. For example, in subjects like medicine, engineering, or law, the accuracy of information can have serious real-world implications. Misunderstanding a medical treatment protocol, a structural engineering principle, or a legal precedent could lead to severe consequences, including harm to individuals. Conversely, in more exploratory or creative disciplines, such as literature or philosophy, the stakes might not be as high, allowing more room for interpretation and discussion around the generated content. Try to think within your course whether there are areas in which the costs of inaccurate information would not be so harmful, and be more encouraging of the use of ChatGPT in these areas.

- **It will get better:** ChatGPT and other AI tools are becoming more accurate and less biased with each iteration. For instance, the advancements from ChatGPT-3.5 to ChatGPT-4 have shown significant strides in reducing errors and enhancing the quality of responses. These improvements are driven by ongoing research, larger and more diverse training datasets, and refined algorithms. While there is no guarantee that future versions of ChatGPT and other AI tools will be more accurate for the sorts of issues you most care about, the general trend is positive.

These ideas cannot eliminate the concerns for inaccuracies or biases but will hopefully help you better think about this challenge, and how to address it in your teaching and more generally in your life.

Concern #3 - Students will become lazy

Another concern is that students will start doing things with ChatGPT that we think they should do on their own, and they will therefore not develop the ability to do these things. A faculty colleague recently told one of us about a practice that he now employs when writing. He writes a paragraph and the first couple of lines of the following paragraph. Then copies and pastes the entire text and asks ChatGPT to finish the second paragraph, which then yields a draft that he can work with. Our colleague then said, "I am worried that ChatGPT is making me lazy as a writer and that I will lose my writing skills." If you have used ChatGPT to repeatedly do a task you used to do on your own, you can probably identify with this colleague. The same holds true for our students.

We think that this risk is very real, just like the appearance of calculators diminished over time the skills that humans have at mental math, or the appearance of GPS made us less able to orient ourselves in an unfamiliar area. When calculators became accessible, they took over the task of complex mental math, making such skills less critical for everyday use. We now continue to teach basic arithmetic in schools because it forms a foundational skill essential for cognitive development and practical application in various professions. However, schools generally allow students to use calculators for most advanced courses or even for basic math calculations in non-math courses like economics. Later in life, the average adult will rely on calculators for most of their day-to-day math needs. Mental math remained important only for specific professions, and this subset of the population had incentives to invest heavily in learning complex mental calculations without a calculator. Similarly for AI, it will be critical to understand what the foundational skills are and how to teach them with or without the use of AI.

In our mind, the key question is which competencies will continue to be relevant in a world where AI is at our disposal, and whether ChatGPT can support our students in acquiring them. Table III.2 below provides a framework we think might be helpful to you in thinking about this challenge.

Table III.2 - Assessment of Competencies

Competencies are...	What To Do
No longer relevant	Drop them
Relevant and ChatGPT can **help** in achieving them	Ask students to use ChatGPT
Relevant and ChatGPT may **hinder** in	Guide students to use ChatGPT in ways

achieving them	that help learning or to not use it at all
Relevant and ChatGPT can **neither** help nor hinder in achieving them	Keep the development of these competencies in your course

Which of the competencies you teach fall in each category? We think the first row (competencies that are no longer relevant) is particularly important. If you think very few of the competencies you seek your students to develop are no longer relevant, we suggest you think a bit more and interrogate how your students are likely to be using each of the competencies after they graduate from your institution. Our sense is that status quo bias will lead many of us to underestimate the number of competencies that fall in this category. We argue that those should be dropped from your course, and be replaced by new competencies that are now relevant for your course and that are perhaps enabled by the use of ChatGPT or a similar tool.

For example, in a writing-intensive course, instead of focusing solely on writing traditional essays, the course could shift towards developing competencies that the job market is increasingly valuing, such as those focused on digital content creation or multimedia storytelling. Additionally, integrating tasks that require collaboration with AI, such as using ChatGPT to draft initial content or brainstorm ideas, might enhance creativity and efficiency in the writing process. We are not writing instructors so if you disagree with our characterization of which type of writing skills will be relevant to your students, we defer to you. But in general it is hard to imagine that, given the ways that ChatGPT and similar tools are reshaping cognitive work, the competencies we want to develop in our students are the same today as 5 years ago.

Ultimately, we would like to nudge you to reassess what you want your students to learn, how AI might help achieve your learning goals, and how you should assess the learning of your students. If you are not sure, perhaps consulting with colleagues who teach in your area of study might be a good way to figure it out.

3. Your Course Policies about Student Use of AI

Given the above, you might feel the need to establish a policy of how your students should use AI tools in your course(s). In this section, we discuss some ideas for policies you could adopt. But before we do so, we would like to note four things that might shape how you think about your policy:

- **Currently, there are no fully reliable ways to detect AI-generated text.** While some AI companies may watermark their text and offer ways to verify if a passage of text was generated by their tool, this will not help you identify if the text was generated by other AI tools.

- **Documenting how AI was used to produce anything will become increasingly difficult.** You might consider asking students to submit their ChatGPT interactions for assignments to evaluate their tool usage (as many educators do, which we'll discuss in Chapter 9). However, as AI is increasingly being integrated into writing platforms like Microsoft Word and Google Docs and even in Apple's operating system, students will receive real-time writing suggestions. This feature allows users to revise text instantly, making it challenging to track how AI was involved in the process.

- **Communicating and enforcing your own policy can be challenging.** Chances are that your institution has a general policy on student use of AI, which might allow you to create your individual course policy. However, if students take multiple courses, and each has a different AI policy, it is likely that you will encounter difficulties in successfully implementing your own specific policy.

- **AI will be available to students once they leave school.** If it aids them in achieving the learning objectives and will continue to do so later in their careers, you need to think carefully about whether prohibiting or regulating its use in your class is helping or hindering your ability to prepare your students for their careers.

Broadly speaking, we see four distinct approaches you may adopt to design your course policy on student use of AI. Their benefits and risks are highlighted in the table below.

Table III.3 - Advantages and Disadvantages of the Four Approaches to Setting an AI Course Policy

Lax	Guided / Flexible	Restrictive	Absent (ignore)
Permit the use of ChatGPT with no restrictions	Permit the use of ChatGPT with guardrails or under specific circumstances	Forbid the use of ChatGPT	Do not set any policy regarding the use of ChatGPT
+ Clear and easy to communicate + Encourages exploration and critical review of AI capabilities	+ Balances the benefits of AI while optimizing the experience from a pedagogical perspective and maintaining academic integrity + Encourages exploration and critical review of AI capabilities	+ Maintains traditional learning methods and standards + Clear and easy to communicate + Eliminates concerns about plagiarism and other detrimental uses of AI	+ Reduces the burden of monitoring and enforcement +Leaves decision-making to student discretion
- Risk of over-reliance on AI for answers - Academic integrity issues	- Complicated to design, communicate and monitor - Students may find rules restrictive or confusing - Difficult to monitor	- Limits exposure to new learning opportunities with AI - Prevents preparing students for an AI-integrated professional environment	- Uncertainty can lead to inconsistent use and fairness (may become in practice the same as a lax policy) - Limits exposure to

- Difficult to monitor	new learning opportunities with AI

Deciding which AI policy to adopt is challenging for several reasons. First, the optimal policy depends on various factors, including the course you teach, your learning goals, and your relationship with your students. Second, technology evolves so rapidly that it's hard to predict which competencies will be important after students graduate. Nonetheless, if you choose to create your own policy, a guided or flexible approach is often the most suitable. Because this is the most pedagogically complex approach, we provide two concrete examples of flexible policies below. The companion site also includes links to AI policies that might interest you.

Example of Flexible Policy #1 - The Traffic Light System

Some instructors have used a system akin to a traffic light to provide guidance to their students on use of AI for different assignments or for different questions within an assignment. The systems vary but by and large the 3 categories represent:

- Green: All forms of AI are allowed
- Yellow: only some forms of AI use are allowed (e.g. brainstorming ideas, receiving feedback, etc.)
- Red: AI is not allowed

Example of Flexible Policy #2 - The Menu System

Danny Liu from the University of Sydney argues in favor of a menu system[11] to guide students on how they may use Gen AI to complete their coursework. He argues that this system provides more flexibility to enable the uses that are best suited to student needs at a particular point in time for a particular assessment. Table III.4 below presents a slightly adapted version of his menu system.

Table III.4 - The Menu System in Detail

Menu Category	Gen AI tasks
As a critical friend	Suggest analyses; Provoke reflection; Provide study/organization tips; Practicing.
Getting started	Suggesting structure; Brainstorming ideas.
Engaging with literature	Suggesting search terms; Performing searches; Summarizing literature; Identifying methodologies; Explaining jargon; Fixing reference lists.
Generating content	Writing some text; Making images, video, and audio; Making slide decks.
Analyses	Performing analyses of data, and text; Suggesting counterarguments.
Editing	Editing tone; Improving clarity and readability; Fixing grammar; Shortening.
Feedback	On all of the above elements; Specifically on rubric criteria.

Each assignment would then have an associated list of the menu items that are allowed for use.

4. What Should You Do?

We hope the above sections will help you make more informed decisions regarding student use of AI in your courses. While we would love to provide you with more concrete guidance, it is difficult for us to do so at this time when everything is changing so quickly. We don't know and are still trying to figure out how to navigate the complex challenges highlighted above. But in the spirit of at least giving you something concrete that you can adopt, tweak, or discard, here is where we land:

1. Because student use of AI can often go undetected, students now have more agency in deciding how much time and effort they devote to doing the work in your course. This raises the stakes in designing assignments that are relevant and in persuading your students that what you are asking them to do is indeed relevant for their lives.

2. If you want to make sure that students do not use AI to do their work, we suggest relying more on the use of assessment methods where students cannot use AI (such as in-person written or oral exams). However, be mindful that these methods may not always accurately measure the competencies your students will need after graduation, especially if their future jobs require using AI technologies. For example, a pen-and-paper statistics exam does not allow instructors to assess whether students can perform statistical analysis on a computer—a crucial skill they will need in the real world.

3. The ubiquity of AI technology calls for redefining educational standards. Ethan Mollick, associate professor of management at Wharton, states in his book "Co-Intelligence"[12] that AI elevates everyone's performance to a minimum level. Bowen and Watson (2024) provocatively argue in their book "Teaching with AI"[13] that AI-generated content should now be the baseline for grading assignments. For example, a clear structure that once earned a B in an essay should now be an F, with higher points for originality and creative expression that require critical thinking and substantial iteration beyond basic AI capabilities.

Whatever you decide to do, we suggest you recognize that AI is here to stay, that your students are using it, and that you are better off devoting some of your pedagogic time and effort to nudging your students to use AI in ways that advance your learning goals. Chapter 8 gives you a sense of how students are using AI to do some of their schoolwork, Chapter 9 suggests ways you can create assignments that ask students to use AI to complete them, and Chapter 10 describes ways you can create bots to leverage AI for student learning.

Chapter #8 - How Students Can Use ChatGPT to Learn

Take a moment to pause and reflect on the mindset that you are bringing to this chapter on student learning. Are you primarily concerned about the potential for increased cheating? Are you skeptical that ChatGPT can contribute positively to the classroom? Are you optimistic that it can serve as a valuable educational tool?

Whether we like it or not, many of our students are already using ChatGPT. Half of postsecondary students in the US are regular users of GenAI, according to a study conducted in Fall 2023.[14] We expect that this percentage will be much higher by the time you read these lines.

In this chapter, we explore some ways in which real students are using ChatGPT in their schoolwork. We believe that, on balance, these uses are more likely to help their learning than to hinder it. At the same time, we recognize that you might disagree with us. Whether you do or not, we hope the examples will broaden your views about how students could use ChatGPT in your courses.

To structure this chapter and the following ones, we use a framework adapted from Mollick and Mollick (2023), who proposed different approaches for student use of AI, along with their pedagogical benefits and risks.[15] For a more in-depth discussion of the pedagogical principles and research that underlie these approaches, we suggest you refer to their seminal paper, also linked on our companion site. The real-world examples we describe don't always fit neatly into a single category, but we think the framework is useful nevertheless to think about students using AI.

Table 8.1 - Ways Students Can Use ChatGPT for Learning

Way	Pedagogical Benefit	Pedagogical Risk
1- Get feedback	* Frequent feedback improves learning outcomes, even if all advice is not taken.	* Not critically examining feedback, which may contain errors.
2- Learn a new topic, concept or skill	* Personalized direct instruction is very effective.	* Risk of inaccurate information. * Serious hallucination risks.
3- Deepen their reflections	* Opportunities for reflection and regulation improve learning outcomes.	* Tone or style of coaching may not match the student. * Risk of incorrect advice.
4- Find arguments and counterarguments	* Alternate viewpoints help deeper learning.	* Hallucination risk and errors.
5- Practice deliberately	* Practicing and applying knowledge aids transfer.	* Hallucination risk.
6- Accomplish tasks	* Helps students accomplish more within the same time frame.	* Outsourcing thinking, rather than work.

Source: Adapted from Mollick and Mollick (2023)

1. Get Feedback

Receiving tailored, high-quality, and timely feedback can have significant benefits for learning. However, providing such feedback to every student is a significant challenge for educators due to the time-intensive nature of the task. ChatGPT can bridge this gap by offering personalized, instant feedback of often reasonable quality. While we do not claim that feedback from ChatGPT is as valuable as the feedback you may provide, we believe it is better than students receiving little to no individualized feedback.

We use the example of Hadar Sachs, a Harvard Kennedy School student, to illustrate the power of using ChatGPT to receive feedback. Hadar needed to prepare a presentation for one of her courses. Generally, she would have practiced this presentation with her partner. However, her partner was not available at the time, so she decided to try out practicing with ChatGPT.

 Hadar Sachs (Harvard Kennedy School Student)

I'm preparing my mini-presentation for dpi200 tomorrow. Here's my class syllabus, the case we are reading this week, and the assignment description and criteria. Only rely on the criteria for the assignment I provide outside of the syllabus document.

[...Uploads syllabus and class reading to chat...]

Each student will prepare a short 5-minute presentation in which they identify a theme/question from the case related to one (or more) of the lenses (power, intersectionality, or ethics) and apply it to their own work experience/background. Students can use whatever media they deem appropriate – (a maximum of 4) slides, (short) video clip, charts/graphics, or talking without visual aids – to support their ideas.

This assignment will be worth 20 points and a submission that receives full credit will:

1. Clearly describe the context of the situation and provide all the details relevant to understand the nuances of your personal case (5 pts).
2. Identify which theme(s) of that class's case relates to the chosen personal example (5 pts).
3. Employ course materials (including readings or case discussions) to discuss the complexities of the personal example (5 pts).

> 4. Name key takeaways from the experience that will inform how you approach similar issues in the future (5 pts).
>
> I am going to say my presentation and I'd like you to give me suggestions to make it clearer, follow the assignment better, and get full points.
>
> Before we start, do you have any questions?

 ChatGPT

No questions from me. Whenever you're ready, please feel free to share your presentation, and I'll do my best to provide feedback based on the assignment criteria and the syllabus.

After giving ChatGPT instructions, including the assignment criteria, Hadar is ready to start reciting her presentation. Using the voice function that was available in ChatGPT 4 version (now also available in ChatGPT 4o), she started recording herself while going through her presentation.

 Hadar Sachs (Harvard Kennedy School Student)

In our case this week, we heard about a crisis in education in the DC public schools and my presentation is going to be about a different crisis, a crisis in the Caribbean back in 2017, when two category 5 hurricanes that look like this hit the U.S. Virgin Islands within two weeks of one another. I had just graduated from college, and for my first job out of college, after I had just graduated with a degree in urban planning from college, and for my first job out of college, I deployed with FEMA to St. Croix, St. Thomas, and St. John. These are the three islands of the U.S. Virgin Islands, which are just southeast of Puerto Rico....

[...rest of her presentation omitted for brevity...]

Please note that what we are seeing in written form in the chat box above is a transcription of Hadar's voice. What's more, if upcoming ChatGPT versions with video capabilities had been available, Hadar could have directly recorded herself in video and asked ChatGPT for advice not only about her content but also about her tone of voice, gestures, etc. For now, we will continue illustrating Hadar's real use of ChatGPT. As she progressed in conducting her presentation, ChatGPT shared reactions and feedback. Below are some of ChatGPT's reactions:

 ChatGPT

That's a compelling introduction, providing a clear context of your deployment to the U.S. Virgin Islands with FEMA. To strengthen your presentation, you might consider specifying which lens (power, intersectionality, or ethics) your experience relates to and how it connects to the themes of the class case. For example, if you choose the lens of power, you could discuss the dynamics of power in disaster response and recovery efforts. If you choose intersectionality, you could talk about how different communities were affected by the hurricanes and the recovery process. If you choose ethics, you could discuss the ethical considerations in disaster response and recovery. What comes next in your presentation?

Hadar continued practicing her presentation several times and recording it with ChatGPT. Something particularly powerful about this example is that Hadar provided a list of criteria for grading the assignment. This allowed ChatGPT to offer tailored feedback on those specific areas, as illustrated below. ChatGPT would not have known anything about Hadar unless she specified it in the prompt. Imagine having to grade a presentation without knowing the assessment criteria, how would you even start? That is why it is critical to provide the key areas or criteria to be assessed in the process of obtaining feedback from ChatGPT.

 ChatGPT

You've provided a detailed account of your experience in community engagement and the challenges faced during the recovery process in the U.S. Virgin Islands. To align your presentation with the assignment criteria, consider explicitly connecting your experience to one of the lenses:

1. **Power:** Discuss the power dynamics between the federal staff, local community leaders, and residents. How did these dynamics impact the recovery process and the prioritization of projects?
2. **Intersectionality:** Reflect on how different communities within the islands may have had varying needs and challenges. How did your team address these diverse needs, and were there any disparities in the recovery process?
3. **Ethics:** Examine the ethical considerations of conducting a community engagement process without guaranteed funding for the outcomes. How did this impact the trust and expectations of the community?

Finally, consider discussing the key takeaways from this experience and how it has informed your approach to similar issues in the future. This will help fulfill the fourth criterion of the assignment.

Once Hadar finished practicing her presentation, she asked ChatGPT to summarize the main points she had made, which she then used as a script during her presentation. As usual, this required a back-and-forth with Hadar providing additional nudging to obtain the output desired.

 Hadar Sachs (Harvard Kennedy School Student)

You missed some of the important details, can you make each set of bullet points tighter but include the details? The bullet points should be from my perspective.

Additionally, she asked ChatGPT to compile all the notes taken into one single script that she could take with her on the day of her presentation. Hadar considered and implemented some of the suggestions received (some were useful, some were not), and then she recited the presentation again using ChatGPT.

Hadar Sachs (Harvard Kennedy School Student)

Here's my second try. Can you give me feedback on the criteria for the assignment?

Hi, everybody. In our case this week,...

[...Full presentation omitted for brevity...]

In this round of feedback, Hadar decided to tailor the feedback even further by asking ChatGPT to adopt the persona of her professor. To do so, she provided details about their work and areas of interest in the next prompt. She could have provided other information she could have considered relevant, such as previous feedback she had received from her professor or relevant papers they may have published related to this topic. Note that she could have asked ChatGPT to adopt multiple other perspectives if it would have been of interest, like that of an economist at the World Bank, a local policy-maker in the U.S. Virgin Islands, etc.

Moreover, Hadar had received mostly positive feedback and encouragement from ChatGPT. Knowing that the model tends to opt for positive reinforcement, she decided to nudge it to be specific about providing areas of improvement.

Hadar Sachs (Harvard Kennedy School Student)

This is my professor: I am an assistant professor of public policy at the Kennedy School of Government at Harvard University. My current research focuses on...

[...description of professor's professional interests and work omitted for brevity...]

Take on his persona and give me 3 critical pieces of feedback.

215

ChatGPT

As your professor, I would like to offer the following critical feedback on your presentation:

- **Clarity and Structure:** While your presentation provides a compelling narrative, it could benefit from a clearer structure that explicitly aligns with the assignment criteria. Consider organizing your presentation into distinct sections that correspond to the assignment's requirements, such as context, theme identification, employment of course materials, and key takeaways. This will make it easier for your audience to follow your argument and understand how each part of your presentation contributes to the overall message.

- **Connection to Course Material:** Your presentation touches on important themes, such as power dynamics and community engagement, which are relevant to our course on politics and ethics. However, I encourage you to more explicitly connect your experiences and lessons learned to specific concepts, theories, or case studies we have covered in class. This will demonstrate your ability to apply course material to real-world situations and deepen the analytical depth of your presentation.

- **Reflection and Future Application:** You've started to discuss some lessons and curiosities from your experience, which is a good start. However, I would like to see a more detailed reflection on how these lessons will inform your future approach to similar challenges. Additionally, consider exploring how your experience and the insights gained might contribute to broader discussions in urban politics and policy, given your background in urban planning and your role in disaster recovery.

Overall, your presentation shows a strong understanding of the complexities of disaster recovery and community engagement. By addressing these areas of feedback, you can enhance the clarity, relevance, and analytical depth of your presentation, making it a more powerful reflection of your learning and experiences.

Hadar wrapped up the dialogue by asking ChatGPT one more time to aggregate her script into one single text with the improvements she wanted to incorporate and to provide a new summary in bullets she could use for her slides. With the time she saved creating a script and improving it, she practiced her presentation one more time, becoming better prepared to deliver it.

2. Learn a New Topic, Concept, or Skill

Learning can be tough. Sometimes students don't have access to explanations or materials that meet them where they are. Other times, the real-life application of what they need to learn is unclear, making it hard to find the motivation to learn. Some students may be embarrassed to admit that they don't understand something. At critical times, instructors may not be available to help or not have sufficient bandwidth for a 1:1 interaction. As described earlier, ChatGPT can act in some ways as a private tutor with endless patience and 24/7 availability, empowering students to take ownership of their learning and drive their progress.

Human tutors add much more value than what ChatGPT can offer, such as inspiring and engaging students or guiding their learning. However, we are hopeful that ChatGPT can be a complement to most educators, particularly those who work with large numbers of students.

In the introduction to Part III of this book, you read about the examples of Sean and Ryan, who learned about risk management using ChatGPT with completely different approaches. In Chapter 10, we will also explore how to design customized chatbots that offer tutoring that is tailored to your instructional needs. All of these examples bring to life the power behind personalizing the learning process with technology.

In this section, we explore in further depth how the agency to choose one's own learning journey can create a deep sense of empowerment. Louis Guerin, a student at the Harvard Kennedy School, learned how to scrape the web (a process of extracting data automatically from websites) and automate an email notification with every new data download by using ChatGPT.

Louis had always thought that web scraping was a really useful skill but had never been able to learn how to do it. As part of one of his assignments in a course about AI, he was asked to automate a process of his choosing. At the time, he was applying for a job in an organization that uploaded new reports frequently on their website and needed to stay up to date on everything they published. He used the occasion to teach himself how to web scrape the reports from the website. Then, he learned to automatically send a notification to his email when a new report was published. His (abridged) exchange is included below:

Louis Guerin (Harvard Kennedy School Student)

Can I tell a Python script to run itself every day?

Notice that when Louis started this exchange, he had never coded in Python and was not even sure of what was possible. He started by checking with ChatGPT if this was feasible. As always, we encourage you to simply skim through ChatGPT's responses unless you are particularly interested in this example.

ChatGPT

Yes, you can set up a Python script to run every day using a task-scheduling tool. On Unix-like operating systems such as Linux and macOS, you can use cron, and on Windows, you can use the Task Scheduler.

Here's how you can set up your script to run every day:

On Unix-like Operating Systems (Linux/macOS):

1. **Edit the cron schedule:** You can schedule a job using cron. Open the crontab file in edit mode using the following command: *[...code omitted for brevity...]*
2. **Add a cron job:** Add a line to the crontab file specifying when you want the script to run. For example, to run your script every day at 2:00 AM, add the following line: *[...code omitted for brevity...]*
   - 0 2 * * * specifies the schedule in the cron format:

- 0: Minutes past the hour (0)
- 2: Hour of the day (2 AM)
- * * *: Day of the month, month, and day of the week (any day)
  - The full path to the Python interpreter and the script should be specified.
3. **Save and exit:** Save the crontab file and exit the editor. Your script will now run at the scheduled time every day.

On Windows:

1. **Open Task Scheduler:** You can find Task Scheduler in the Start menu or by searching for it.
2. **Create a new task:** In Task Scheduler, click on "Create Task" in the right sidebar.
3. **Configure General Settings:** In the "General" tab, give the task a name and description.
4. **Configure Triggers:** In the "Triggers" tab, click "New" and set the schedule for the task (e.g., daily at 2:00 AM).
5. **Configure Actions:** In the "Actions" tab, click "New" and select "Start a program." In the "Program/script" box, enter the path to your Python interpreter (e.g., C:\Python39\python.exe) and in the "Add arguments" box, enter the path to your script (e.g., C:\path\to\script.py).
6. **Configure other settings:** Adjust other settings (e.g., "Conditions" and "Settings" tabs) as needed.
7. **Save the task:** Once you have configured everything, save the task.

With either of these methods, your script will run automatically at the scheduled time every day. Make sure to provide full paths to your Python interpreter and script file in the scheduler.

ChatGPT not only confirmed that it is possible to run a script every day on Python but also provided detailed instructions on how to make it happen. You may be thinking that he could have found a blog on a Google search explaining the same process, but the power of ChatGPT comes in its ability to interact with us. If any of these steps felt overwhelming or unclear, Louis could have followed up to ask clarifying questions.

Now that Louis realized that this was possible, he invested some time in planning how to approach the task. He could have just asked ChatGPT: "Give me a Python code that notifies me every time a new report is uploaded to this website". However, he had learned that complex tasks are better accomplished when they are broken down into smaller steps (as we suggested in Chapter 2). Knowing this, Louis landed on creating a Python script that would first web scrape the report, add the link to the new report in a Google Sheet, and then another script that would notify him daily of any new additions. Louis was able to break the task down into simpler steps thanks to some experience he had creating macros in Excel, but he admitted a student with no experience at all with automation may have struggled with this step. Let's see how his interaction with ChatGPT continued.

 Louis Guerin (Harvard Kennedy School Student)

I have this code, and want it to send me an email if the final condition is met. Can you help me?

 ChatGPT

Absolutely, I can help you. In order to send an email based on a condition in your code, you can integrate an email-sending functionality at the appropriate point in your program.

To send an email in Python, you can use the smtplib library. Here is a brief example of how to use smtplib to send an email. You will need to replace the placeholder values with your own information (email provider's SMTP server and port, your email, password, recipient's email, subject, and body of the email):

Here's what you need to do:

- Make sure you have the smtplib library installed. It should come pre-installed with Python, but you can check with import smtplib.
- Replace "your_email@gmail.com" with your email address and "your_email_password" with your email password.
- Replace "recipient_email@gmail.com" with the email address where you want to send the notification.
- Replace "Condition met" with the desired subject of your email and "The final condition has been met in the program." with the desired body of your email.
- Place the send_email() function call inside the conditional block of your code where the final condition is met.

Keep in mind that some email providers (e.g., Gmail) may require you to enable access for less secure apps in your account settings or generate an app-specific password to allow sending emails through their SMTP servers. Be sure to check the settings of your email provider for more details.

Louis' conversation with ChatGPT did not end there. He had several back-and-forth exchanges in which he asked ChatGPT to improve its code, or reported errors he was obtaining and sought help to troubleshoot them. Below are some of the additional prompts he sent:

- I get "NameError: name 'googleapiclient' is not defined.
- I got this "Request had insufficient authentication scopes".
- Check Scope: Ensure that your JSON key file has the scope 'https://www.googleapis.com/auth/gmail.send' included. Your script seems to be already using this scope, but double-check that it is authorized in our Google Cloud project. How do I do this?
- Is there a way to get Google Drive to send me a notification when a file is changed?

He approached this learning process with an experimental mindset, not aiming for the perfect prompt in every interaction but rather being ready to iterate and tackle the same problem through multiple routes. He estimated it took him about six hours to accomplish this task, which could have otherwise taken days or more. What kept him motivated despite the challenges was the initial progress he made and the excitement about the potential to automate the task. Now he knows that if he ever needs to scrape the web or do a similar task requiring coding, he will be able to figure out how to do it. He shared: "I had never coded before and never thought I could do it – using an LLM to learn that skill was incredibly empowering."

3. Deepen Their Reflections

ChatGPT can increase metacognition (the ability to be aware and understand one's own thought processes) by helping students articulate their thoughts, provide alternative viewpoints, or deepen their reflections. Mollick and Mollick (2023) refer to this as using ChatGPT as a coach. They suggest an example prompt that can be used by students in this process and that is also included in our companion site.

Below, we illustrate how this may look in practice with an example from Blake, a hypothetical student who wishes to engage in a reflection process after finishing a team project.

Blake

Role: You are a helpful friendly coach helping me reflect on a recent team experience. Introduce yourself. Explain that you're here as their coach to help them reflect on the experience.

Instructions: Think step-by-step and wait for me to answer before doing anything else. Do not share your plan. Reflect on each step of the conversation and then decide what to do next. Ask only 1 question at a time. 1. Ask me to think about the experience and name 1 challenge that we overcame and 1 challenge that we did not overcome. Wait for a response.

Constraints: Do not proceed until you get a response because you'll need to adapt your next question based on my response. 2. Then ask me: Reflect on these challenges. How has your understanding of yourself as a team member changed? What new insights did you gain? Do not proceed until you get a response. Do not share your plan with students. Always wait for a response but do not tell me you are waiting for a response.

Pedagogy: Ask open-ended questions but only ask them one at a time. Push me to give you extensive responses articulating key ideas. Ask follow-up questions. For instance, if I say I gained a new understanding of team inertia or leadership, ask me to explain my old and new understanding. Ask me what led to my new insight. These questions prompt a deeper reflection. Push for specific examples. For example, if I say my view has changed about how to lead, ask me to provide a concrete example from my experience in the game that illustrates the change. Specific examples anchor reflections in real learning moments.

Personalization: Discuss obstacles. Ask me to consider what obstacles or doubts I still face in applying a skill. Discuss strategies for overcoming these obstacles. This helps turn reflections into goal setting.

Pedagogy: Wrap up the conversation by praising reflective thinking. Let me know when my reflections are especially thoughtful or demonstrate progress. Let me know if my reflections reveal a change or growth in thinking.

ChatGPT would respond to this prompt by nudging Blake to think deeper about their team experience, asking them questions that make them consider rich reflections.

4. Find Arguments and Counterarguments

Creating a presentation or essay is a common type of assignment that students face. ChatGPT can help with the initial steps of finding a thesis and deciding on a structure or outline. We consider these uses of ChatGPT to be examples of "accomplishing a task" (covered in section #6 of this chapter). The next step, finding arguments and counterarguments to a thesis, we believe has great potential for learning enhancement. The process of coming up with arguments and counterarguments to a thesis is critical to learning to consider different perspectives, understand a topic in-depth, and defend robust and credible ideas.

We will illustrate ChatGPT's ability in this area with an example from Angela Pérez, one of the authors of this book and a recent graduate from the MPA in International Development program at Harvard Kennedy School. Angela was writing an essay for her philosophy class about the idea of morality of Peter Singer, Emeritus professor of bioethics at Princeton University. She could have taken an easy route by simply asking ChatGPT for possible arguments and counterarguments to a given thesis. We consider this to be the "lazy use" of ChatGPT. An example of a simple prompt is included below.

 Angela Pérez (Harvard Kennedy School Student) [Simple prompt]

I want to write an essay that defends the following moral argument:

"When we buy new clothes not to keep ourselves warm but to look 'well-dressed' we are not providing for any important need. We would not be sacrificing anything significant if we were to continue to wear our old clothes and give the money to famine relief. By doing so, we would be preventing another person from starving. It follows from what I have said earlier that we ought to give money away, rather than spend it on clothes which we do not need to keep us warm. To do so is not charitable, or generous. Nor is it the kind of act which philosophers and theologians have called 'supererogatory'-an act which it would be good to do, but not wrong not to do. On the contrary, we ought to give the money away, and it is wrong not to do so. ... the present way of drawing the distinction, which makes it an act of charity for a man living at the level of affluence which most people in the 'developed nations' enjoy giving money to save someone else from starvation, cannot be supported."

Please provide 5 arguments in favor of this thesis, and 5 arguments against this thesis.

With this prompt, Angela would be merely a recipient of information; she would not be actively engaging in the effort of coming up with different perspectives. However, Angela could still critically assess the strengths, relevance, and validity of the arguments and counterarguments suggested by ChatGPT. In doing so, Angela would be blurring the lines between one use of ChatGPT that potentially hinders learning (by shortcutting the process of coming up with different perspectives) and one that potentially enhances learning (by presenting information that needs to be judged and critically assessed).

Note that as Angela's instructor, you would have no way of knowing whether Angela just engaged in the "lazy use" of ChatGPT (replicating the arguments and counterarguments received) or whether she critically assessed the arguments and counterarguments suggested by ChatGPT (hopefully gaining some learning in the process). Nonetheless, this might not be that much different from the previous status quo. Without ChatGPT, Angela could have done a literature review and scanned different papers in search of the arguments and counterarguments to her thesis. Maybe Angela could come up with some arguments on her own, but maybe she would have found papers that cover the same or similar research questions and simply replicate what an existing body of literature argues. ChatGPT has made this job much easier for Angela to engage in the "lazy way", but the opportunities to skip the step of critically engaging with the arguments in favor and against a thesis are not just appearing now.

An alternative approach to this task that we believe is learning enhancing consists of doing a roleplay with ChatGPT. This is, Angela could ask ChatGPT to adopt a specific persona and to have a back-and-forth debate with her about a given idea. This way of using ChatGPT in the process of finding arguments and counterarguments preserves the principle of active learning covered in Chapter 2. This is exactly what Angela did. She chose this route because she found the "lazy way" to be very superficial, and she believed this approach would help her address specific thoughts or questions that she had pondered about the topic she was writing about.

In the prompt below, Angela asked ChatGPT to:

- Adopt the persona of a philosopher who is enthusiastic about Peter Singer's moral philosophy.
- Have a debate with her around the validity of a moral argument she specified in the prompt.
- Assess her understanding of the key trade-offs inherent in Peter Singer's moral theory.

 Angela Pérez (Harvard Kennedy School Student)

I'm a student in a graduate public policy program trying to learn about Peter Singer's philosophy, as proposed in "Famine, Affluence and Morality". To help deepen my understanding, I want to simulate a conversation between two philosophers debating the following moral argument:

"When we buy new clothes not to keep ourselves warm but to look 'well-dressed' we are not providing for any important need. We would not be sacrificing anything significant if we were to continue to wear our old clothes and give the money to famine relief. By doing so, we would be preventing another person from starving. It follows from what I have said earlier that we ought to give money away, rather than spend it on clothes which we do not need to keep us warm. To do so is not charitable, or generous. Nor is it the kind of act which philosophers and theologians have called 'supererogatory'-an act which it would be good to do, but not wrong not to do. On the contrary, we ought to give the money away, and it is wrong not to do so. ... the present way of drawing the distinction, which makes it an act of charity for a man living at the level of affluence which most people in the 'developed nations' enjoy giving money to save someone else from starvation, cannot be supported."

You will play a philosopher who is enthusiastic about Singer's moral philosophy and you are trying to convince me of the moral validity of the above argument, though you are also curious and amenable to changing your mind. You have a strong understanding of Singer's philosophical thinking and its implications.

Your ultimate goal is to use this conversation as a way to assess my understanding of the key tradeoffs inherent in adopting Singer's philosophical view. Keep the conversation

Note that if Angela had not been satisfied with the level of competence that ChatGPT displayed with Singer's philosophy, she could have experimented with more detailed instructions like mentioning explicitly his books or seminal articles to ensure ChatGPT fed its responses considering these materials.

 ChatGPT

I understand your request. Now, let's begin our conversation:

As a proponent of Peter Singer's moral philosophy, I find his arguments in "Famine, Affluence, and Morality" particularly compelling. He challenges us to reconsider our ethical responsibilities towards others, especially in the context of global poverty and suffering. His argument about spending money on non-essential items like clothes, when the same resources could prevent significant suffering, raises profound ethical questions.

Singer asserts that if we can prevent something bad from happening, without sacrificing anything of comparable moral importance, we ought, morally, to do it. Purchasing clothes for reasons beyond basic necessity, while others suffer from extreme poverty, seems to contradict this moral imperative.

What are your thoughts on this perspective? Do you think that our moral responsibility extends to the point of forgoing non-essential luxuries in favor of aiding those in dire need?

In its response above, ChatGPT fully adopts the persona and simulates a conversation by inviting Angela to share her views on a specific question. Angela had read Peter Singer's books and had some thoughts on the question. She proceeded to share them as captured below. Notice that in this process, Angela had to be creative about finding flaws in ChatGPT's logic, offering counterarguments, and thinking critically about examples of her own life that may support or weaken a given argument.

 Angela Pérez (Harvard Kennedy School Student)

I find that there are various flaws in this logic:

(1) He assumes that we can prevent something bad from happening with those same funds, which may not always be true because even if we live in a global world, there are information asymmetries that prevent people in the West to know how to donate their money so it's used to reduce suffering (e.g. how to know which NGO to donate to? Many NGOs are not effective, how to judge which ones to donate to, in order to ensure the money is used effectively?)

(2) He assumes we can fairly straightforwardly judge what are non-essential items. While his argument is obvious and I would agree fancy or luxurious clothes are morally wrong, it is not always crystal clear. Here are a few examples. Example 1: individuals in developed countries struggle to assess how much they should save for retirement, and therefore what is an essential level of saving (say, to cover medical costs, retirement life...) and what is non-essential and could be donated. Example 2: luxury goods may sometimes be required for people trying to maximize their impact across their lives. So if an investment banker finds that the most good they can do is through being an investment banker and making a lot of money to donate afterward, that investment banker, to promote and do well in their career and make the most amount of money, will need to network, and travel to fancy places with colleagues and clients and wear fancy clothes to be well regarded in the industry, etc.

(3) It could be argued that affluent people have no moral obligation to prevent significant suffering because they have earned their money through hard work. If we all across society adopted this philosophy, the incentive system in which the functioning of our economy rests (say, no one would decide to work hard and earn more money and progress in their careers because that would have no monetary rewards), would be distorted and its impact on the overall economy is unclearly beneficial

Now Angela has exposed all her ideas and reactions to the prompt. It would be difficult and time-consuming for any educator to engage with each of her arguments and provide reactions and detailed feedback on them. Even more to do this with each student. However, ChatGPT can help (even if it might not be able to do it at the same level as the educator).

 ChatGPT

You've raised some important points that are worth exploring in the context of Singer's philosophy. Let's address each of your concerns:

- **Effectiveness of Charitable Giving:** You're right to point out the challenge of information asymmetry in charitable giving. Indeed, one of Singer's implicit assumptions is that we can effectively prevent suffering with our donations. However, Singer might argue that the existence of some ineffective charities doesn't negate our moral obligation to help; instead, it calls for due diligence in finding effective ones. Organizations like GiveWell have emerged precisely to address this concern, helping donors identify the most effective charities. The key here is the intent and effort to ensure that our donations do as much good as possible.

- **Defining Non-Essential Items:** This is indeed a complex issue. Singer's argument relies on a somewhat utilitarian view, where the moral action is the one that results in the greatest good for the greatest number. The examples you've given, like saving for retirement or the investment banker's career requirements, demonstrate the nuanced nature of what can be considered 'essential'. Singer might suggest that while some level of personal and career investment is necessary, there's often a point beyond which additional spending serves more to increase personal comfort or status rather than contributing to essential needs. The challenge, then, is to honestly assess where that line is in our own lives.

- **Moral Obligation and Economic Incentives:** The argument that affluent people

---

have no moral obligation to prevent significant suffering because they earned their money through hard work is a common one. Singer, however, would likely argue that our moral obligations extend beyond our immediate circles and that the accident of being born into affluence doesn't diminish our responsibility to others. Regarding economic incentives, Singer might suggest that a shift in societal values towards greater empathy and altruism need not undermine economic productivity. Instead, it could lead to a more equitable and sustainable economic system.

In conclusion, while Singer's views challenge deeply ingrained notions of personal entitlement and societal norms, they also invite us to reevaluate our priorities and the impact of our actions on a global scale. What are your thoughts on these points? Do you think there's a balance to be struck between personal economic incentives and moral obligations to others?

ChatGPT addressed each of Angela's concerns and offered back new questions, inviting her to a dialogue. Angela continued the dialogue with new interactions. Some of ChatGPT's viewpoints she considered more helpful, and some others she considered weak. It is helpful that she was familiar with Singer's philosophy, as otherwise, it would have been harder for her to critically assess some of ChatGPT's claims. Sometimes she answered with beliefs of her own, and other times she answered with claims she had heard from peers she had debated with. Her goal was to obtain an overall picture of the trade-offs in Singer's moral philosophy. Here are some of her subsequent prompts as she continued the dialogue:

- Could you say more about what Singer would say about the shifting of societal values leading to a more equitable and sustainable economic system?
- I am not convinced. Are you saying that the current level of wealth in the economy is enough to reduce all suffering if it is appropriately redistributed? And that, people would still be motivated to be productive and generate further wealth and progress, even if they cannot benefit from the product of their efforts beyond a very basic level? That sounds pretty removed from reality and from human nature. Also, it may be

true that affluent nations/individuals can do good with their current level of wealth, but they cannot reduce all suffering; most of the suffering in the world is a result of politics, war, conflict... and money cannot do anything about that.

- But why is it wrong to begin with to not give money away and to buy a fancy sweater? What if it makes someone happy to pamper themselves or if it helps them reflect their identity to dress extravagantly?
- How would we evaluate for example, in this framework, buying a sweater that is more expensive but that it's of higher quality and more durable through time, vs. buying a sweater that is cheaper and allows me to donate, but may be damaged in a year and therefore I'll need to replace it buying a new one?

After a while, she hit "Finished" and received ChatGPT's overall assessment of her interventions in the debate.

 ChatGPT

Great, let's assess your understanding of the key trade-offs inherent in adopting Peter Singer's philosophical view:

1. **Prioritizing Global Needs Over Personal Luxuries:** You've recognized the core of Singer's argument, which is about re-evaluating our spending on non-essential items in light of the pressing needs of those suffering from poverty and preventable diseases. You understand that this involves a significant shift in how we view our moral obligations and personal consumption.
2. **Challenges in Effective Altruism:** You've insightfully pointed out the practical difficulties in ensuring that donations are used effectively and the complexities in defining what constitutes a 'necessary' expense. This demonstrates your grasp of the nuanced nature of applying Singer's philosophy in real-world scenarios.
3. **The Role of Economic Incentives:** Your skepticism about the balance between altruism and personal economic incentives reflects an understanding of the

potential tension between Singer's ideals and the realities of human motivation and societal structures.

4. **Complexity in Ethical Decision-Making:** Your questions about purchasing decisions, like the choice between a durable versus a cheaper sweater, show an appreciation for the complexity of ethical choices in everyday life, considering both long-term sustainability and immediate charitable impact.

Overall, you've demonstrated a thoughtful engagement with Singer's philosophy, understanding its fundamental principles while also critically considering its practical implications and challenges. You've navigated the key trade-offs well, showing an awareness of the balance between ethical ideals and real-world complexities.

Do you have any further questions or topics you'd like to explore in this area?

Realizing that ChatGPT offered mostly positive feedback (as Hadar did in her example above), Angela made sure to prompt for specific areas of improvement.

 Angela Pérez (Harvard Kennedy School Student)

What are key trade-offs or weaknesses in Singer's argument that I have not uncovered? Please share at least 5 of them.

After reviewing the response from ChatGPT (omitted for brevity), which highlighted her blind spots in her knowledge or perspective, she asked ChatGPT to provide a summary of the key trade-offs of Singer's morality argument. She then leveraged this summary in building her essay.

 Angela Pérez (Harvard Kennedy School Student)

Please review all our previous conversation, and provide a table where you evaluate the moral argument I shared in the beginning. Please provide three columns: (1) Key ideas from the moral argument (2) Supporting argument from Singer's philosophy (3) Counterargument. Please do this in less than 500 words.

The final table created (omitted for brevity) still needed refinement and adjustments. However, Angela felt that this roleplay helped her arrive at a level of depth and engagement with specific ideas that she rarely reaches when she is writing essays on her own!

5. Practice Deliberately

Hands-on practice is key for effective learning. Yet students don't always have opportunities to practice and apply their knowledge. Sometimes there are not many practice problems available (partly because creating them is time-consuming for teachers). Other times, students have no means of receiving feedback to ensure they are practicing correctly. ChatGPT can help students by creating practice problems with solutions.

This was the case of Danielle Berger, a student at Tufts University. Danielle was preparing for the final exam of her Biology class on Cells and Organisms. This was a multiple-choice type of exam. Danielle's professor had posted a list of incomplete questions and answers for students to get a sense of the type of questions that could come up in the exam. For example, one sample question would be redacted as follows: "What is a possible subsequent chemical process in which [compound above] might be a reactant?" with no answer key.

Danielle felt that these sample questions were not enough for her to practice on her own and ensure she was ready for the exam. With the help of ChatGPT, she brainstormed potential exam questions and quizzed herself to assess her understanding of the lecture. Her conversation is captured below.

 Danielle Berger (Tufts University Student)

233

Which of the following best describes the conversion of [a reactant to a product shown above]?

Give me an example of a question with four answer choices

 ChatGPT

I'd be happy to provide an example of a completed question for you:

Question: "Which of the following best describes the conversion of glucose to pyruvate during glycolysis?"

Answer Choices:

    a.   Oxidation-reduction reaction
    b.   Hydrolysis reaction
    c.   Decarboxylation reaction
    d.   Esterification reaction

In this example, the question describes a specific biochemical process (the conversion of glucose to pyruvate during glycolysis), and the answer choices include various types of chemical reactions. The correct answer in this case would be "a. Oxidation-reduction reaction," as glycolysis involves multiple oxidation and reduction steps in the conversion of glucose to pyruvate.

Please note how ChatGPT turned a generic question into a specific one and created four different alternative answers. ChatGPT also shared the right solution to the question. To further improve her learning process, Danielle could have asked ChatGPT to not share the solution and to quiz her instead.

Part of the beauty of ChatGPT is that it can produce an endless number of questions in a short amount of time. In her next interaction, Danielle asked ChatGPT to continue providing different test questions. She practiced with about 20 different questions produced by ChatGPT. Here is another example.

 Danielle Berger (Tufts University Student)

Treatment with an herbicide (molecule or reagent that kills plants) results in the disruption of X. Which of the following is a possible mechanism by which this herbicide works?

Give me an example of a question with four answer choices redacted.

ChatGPT

Certainly, here's a sample question based on your provided text:

Question: "Treatment with an herbicide (molecule or reagent that kills plants) results in the disruption of [redacted]. Which of the following is a possible mechanism by which this herbicide works?"

Answer Choices:

a. Inhibition of photosynthesis
b. Suppression of root growth
c. Promotion of flowering
d. Enhancement of water uptake

In this question, you are asked to consider the potential mechanism by which an herbicide works, specifically regarding the disruption of a certain process (indicated by the redacted portion). The answer choices provide various mechanisms that herbicides can employ, but the correct answer would depend on the specific effect or process targeted by the herbicide.

After seeing this question, Danielle was not clear about the solution's explanation, so she asked a follow-up clarifying question: "If Photosystem II is inhibited how will photosynthesis react?" ChatGPT responded with a full explanation of the effects of inhibiting Photosystem II on photosynthesis.

As she was practicing for her exam, Danielle was aware that she could not fully trust the explanations from ChatGPT and checked her textbook and readings when she wanted to double-check a specific explanation. She faced two main risks that we have discussed elsewhere. The first one is the risk of hallucination, this is, ChatGPT providing incorrect answers to the questions. The second one is that the questions suggested by ChatGPT may not be tailored to her course (e.g. some may be too advanced or not focused on the main learning goals of the course). Chapter 10 will explore how you as an educator can reduce these risks by creating custom chatbots for your students.

6. Accomplish Tasks

ChatGPT can be a powerful productivity tool, helping students accomplish tasks and freeing up time that they can decide to spend in other ways (at times, potentially more valuable for learning). Some of the tasks that students can accomplish with ChatGPT include:

- Creating an outline or structuring ideas.
- Supporting or performing data analysis and visualizations.
- Summarizing a text passage.
- Taking notes.
- Automating tasks.

We explore these uses of ChatGPT below.

Creating an outline or structuring ideas

We start by illustrating an example of how a student may go about the process of writing an essay. As captured in section 4 above, this process involves multiple steps, including doing research, finding a thesis, creating an outline, structuring the arguments, and finding data to support those arguments. Students may use ChatGPT to do everything for them (e.g., "write an essay on the upcoming U.S. elections"), to help them on one of the stages (e.g., "suggest an outline that I can use in an essay about the upcoming U.S. elections"), or as a partner that they rely on in a scaffolded process of building out the outline.

We shared in section 4 an example from Angela where she used ChatGPT to find arguments and counterarguments to a thesis. This use, we argued, enhanced her learning. The two examples we include below give you a flavor of other ways in which students can leverage ChatGPT when writing essays. These are more focused on the benefits of automation and time saving than on the benefits of improved learning.

Let's start with an example from Angela. She was preparing to write an essay for her Political Philosophy for Development class. In her initial prompt (captured below), Angela provided the question that she needed to respond to, followed by context about her own interests and professional experience. Even though she had some ideas about which topic she wanted to write about, she decided to start with a brainstorming session. This would allow her to see if any other ideas would be more interesting to explore from a philosophical point of view.

 Angela Pérez (Harvard Kennedy School Student)

I need to answer this question for an essay: "Take a real-world policy or applied problem (perhaps one that you have worked on in your career) that we didn't tackle in class and identify (a) what philosophical questions this policy or problem raises or connects with, (b) how different political philosophies (Rawlsian, utilitarian, capabilities, etc.) would treat them, and (c) what view you would hold based on careful philosophical argument. " I have previously worked in education, youth unemployment, and trade policy in Africa. I am from Spain. What policy questions would be interesting for me to explore in this essay?

Please suggest at least 10 topics I could explore, along with 3 bullet points that capture why these topics would be interesting from a philosophical point of view.

 ChatGPT

To answer this question step-by-step, I will first suggest 10 policy topics that you could explore for your essay. After that, I will provide three bullet points for each topic, explaining why they are interesting from a philosophical point of view.

1. Education Policy in Africa:

- Philosophical Questions:

  - What is the moral obligation of governments regarding access to education?

  - How does education contribute to social justice and equity?

  - Is there a universal model of education that respects cultural diversity?

2. Youth Unemployment Programs:

- Philosophical Questions:

  - How does youth unemployment relate to distributive justice?

  - Is it ethical to prioritize youth employment over other age groups?

  - What role does individual agency play in addressing unemployment?

3. Trade Policy in Africa:

- Philosophical Questions:

  - What ethical considerations should guide international trade agreements?

  - How does trade policy impact economic development and global inequality?

  - Is there a duty to protect domestic industries or promote global free trade?

4. Non-profit Governance:

- Philosophical Questions:

  - What are the moral responsibilities of nonprofit organizations?

  - How should nonprofits balance their mission with financial sustainability?

> - Is there an ethical framework for evaluating nonprofit impact?
>
> *[...selected content omitted for brevity...]*
>
> Now that I've provided the 10 policy topics and their respective philosophical questions, you can choose the one that resonates with you the most based on your experiences and interests. Once you've made your selection, you can delve deeper into analyzing the philosophical aspects, contrasting political philosophies, and articulating your own view for your essay.

The breadth of policy topics was useful in getting Angela to consider new directions in which she could take her essay. However, the specific topics suggested by ChatGPT were too broad. Let's see how she addresses this in the next interaction. In the below prompt, she tried to give new instructions to ChatGPT by providing an example of what she considered to be a specific enough topic. She included a question that she had been wondering about during her professional experience in Africa to nudge ChatGPT to arrive at the level of specificity she desired.

 Angela Pérez (Harvard Kennedy School Student)

> These are good questions, but they are very broad. Can you think of 10 more specific topics, and then give me 3 reasons why these topics would be interesting from a philosophical perspective? The topics should be something as specific as: "deciding to spend resources educating the African elite vs. educating the majority of the population"

ChatGPT provided a new set of questions that were more specific than the initial round, but they were still too broad for Angela's liking. Despite this, the process helped her transition from the intimidating blank page to generating a few intriguing philosophical questions. She started considering a tension in donor-recipient relationships as the topic for her thesis. Angela then checked with ChatGPT to see if this would be a suitable topic for her essay. ChatGPT's answer is captured below.

ChatGPT

The tension in donor-recipient relationships, particularly the dynamic between obligation for oversight versus the imposition of burdensome reporting requirements, is indeed a philosophically rich topic to explore in response to your essay prompt. Here's how it aligns with the different aspects of the prompt:

(a) Philosophical Questions Raised

1. Ethics of Power and Control: The issue raises questions about the ethical use of power by donors. How should power dynamics between donors and recipients be managed to avoid exploitation or paternalism?
2. Justice and Fairness: It questions the fairness in the relationship—do burdensome requirements unfairly disadvantage recipients, possibly overshadowing the benefits of the funding?
3. Autonomy vs. Accountability: This tension also highlights the philosophical conflict between granting autonomy to recipients to manage their affairs and ensuring accountability for the resources provided.

(b) Treatment by Different Political Philosophies

1. Rawlsian Philosophy (Theory of Justice): From a Rawlsian perspective, the focus would be on ensuring that the arrangements between donors and recipients do not disadvantage the least well-off, in this case, potentially the recipients of the aid. Policies should be structured to ensure fairness and equitable opportunities for all parties to benefit from the arrangement.
2. Utilitarianism: A utilitarian approach would evaluate these relationships based on the overall happiness or utility produced by the donor-recipient interaction. If stringent oversight and reporting requirements maximize overall utility by ensuring effective use of funds, they might be justified, even if they are cumbersome.
3. Capabilities Approach (Sen and Nussbaum): This perspective would focus on whether the funding and its associated conditions genuinely enhance the capabilities of the recipients to achieve their objectives. The approach would

criticize practices that, though well-intentioned, effectively hinder recipients' ability to act and meet their community's needs.

(c) Personal Philosophical View

Based on these considerations, your personal view might synthesize elements from these philosophies. For instance, you could argue that:

- Balanced Approach Advocacy: While oversight is necessary to ensure responsible use of funds, it should not compromise the recipients' autonomy and effectiveness. A balance must be struck that respects the dignity and capabilities of recipients, aligns with principles of justice by treating them fairly, and still satisfies the utilitarian aim of maximizing overall outcomes.
- Ethical Reciprocity: You might advocate for a relationship that is more of a partnership, where both donors and recipients agree on what constitutes meaningful impact and reasonable accountability measures, emphasizing mutual respect and understanding.

This topic not only allows for an exploration of the specific ethical tensions in donor-recipient dynamics but also connects these to broader philosophical debates about justice, autonomy, and the ethical distribution of power and responsibility. Thus, it offers a rich field for philosophical analysis and argumentation in your essay.

Angela read through the answer and was satisfied with the potential of the question. ChatGPT allowed her to select an interesting topic and provided her with multiple questions or topics she could explore in her essay, in less time and in a more enjoyable way than if she had done this alone. However, notice how she did not just take the first output that ChatGPT provided; she refined and iterated it so she could explore a question that was interesting to her and that fit the question provided by the professor. She was able to do this because she was familiar with the topic and had some sense of what a great philosophy essay would look like. A less experienced student, someone who is not interested in investing time in this activity, or someone who is facing their first philosophy essay may not have had these assets to guide their interactions with ChatGPT.

Angela's final essay only partially followed the outline that ChatGPT had provided. She added her own opinions on the matter, incorporated information about the shift in the field to looking for tangible impact, included specific content, vocabulary, and principles that had been discussed in her Philosophy class, and elaborated on issues that she thought were crucial. But ChatGPT helped her get started more productively, choosing a more interesting topic and structuring her thoughts around the issue.

Now we turn to an example from Sofia Aron, a Harvard Kennedy School student who was writing a reflection paper for a class called "Public Crisis." These reflection papers were weekly assignments on various topics covered during class. Midway through the semester, she decided to experiment with how she could use ChatGPT to help her finalize these essays more efficiently. Her interaction with ChatGPT started by setting the stage for how she would use the conversation to support her in this process.

 Sofia Aron (Harvard Kennedy School Student)

I will use this chat so we can iterate ideas about my assignments in the class: MLD 385 Public Crisis.

I will first upload my previous reflections so that you can have an idea of my writing style and what's expected

 ChatGPT

Absolutely, I'd be glad to help you brainstorm and refine your ideas for your assignments in MLD 385 Public Crisis. Whether you need assistance with research, developing concepts, structuring your work, or understanding key principles related to public crises, feel free to share the details of your assignment, and we can start from there.

> Please go ahead and upload your previous reflections. Once I have a look at them, I can get a better sense of your writing style and the expectations for your assignments. This will help me provide you with more targeted assistance.

In the next interaction, Sofia uploaded her six reflection papers from the past weeks and asked ChatGPT to read them and analyze her writing style. This way, she would ensure that the model would consider her writing style when suggesting ideas for the essay.

Now Sofia was ready to start preparing for her next reflection paper. She had some ideas of her own on the topic after hearing that week's lecture and considering some of her own professional experience. These were rough drafts or ideas, and she sought ChatGPT's help to structure them and build on them to flesh out a one-page reflection paper.

 Sofia Aron (Harvard Kennedy School Student)

Now help me prepare a one-page reflection on when a private sector issue should be made public.

These are some of the ideas I already have. Please improve and complement taking into consideration my writing style

To answer this question I will first like to define what "making an issue public" means for the objective of this reflection. Making an issue public would mean making information available and transparent about the issue to the public about the issue so that there is a public debate among different stakeholders (government, citizens, NGO, private sector, politicians, media outlets, etc.).

Given this definition, I can think of the following factors:

- **When the public is negatively impacted:** if an issue related to the private sector is having a negative impact on a certain population, then this should be public. We see

> many private sector companies make it public whenever they are having a positive impact. For example, if a private company realizes that one of its products could improve the probability of a disease, then this should be discussed with the public so that the negative impact is reduced.
>
> - **Accountability is needed:** when holding the issue only among private parties will create a negative incentive to not act upon the issue accordingly and thus public accountability is needed to ensure the issue is treated correctly. For example, recently in Peru, there was an oil spill caused by a private company, since some information has not been disclosed to the public by the private company like for example the amount of contamination, the company has been able to reduce their cleaning efforts. If this information had been disclosed since the beginning, then different stakeholders like the population of the area or media outlets could have generated more pressure.

Notice how Sofia asked ChatGPT to take into consideration her writing style. While many have criticized ChatGPT's writing as mechanical and artificial, it is possible with some prompting to change the tone and style of ChatGPT's responses. ChatGPT's response is included below.

 ChatGPT

Based on your ideas and writing style, I'll craft a one-page reflection that incorporates and expands upon your initial thoughts. Given the analytical and precise nature of your previous reflections, I'll aim to maintain clarity and directness in the writing. Here's how we could structure the reflection:

Introduction

- Begin with the definition of "making an issue public" in the context of the reflection.
- State the importance of public discourse on private sector issues.

Factors for Making an Issue Public

- Elaborate on the significance of the public impact, using the example of a private company's product affecting health.
- Discuss the need for accountability, referencing the oil spill case in Peru as an example of the consequences of withholding information.
- Mention national security issues as a clear criterion for public disclosure.
- Suggest additional factors such as legal obligations and ethical considerations.

Conclusion

- Summarize the key points.
- Emphasize the balance between privacy and transparency for the public good.

Here's a draft of the reflection:
[...draft omitted for brevity...]

In the excerpt above, ChatGPT suggested an outline as well as a full draft based on Sofia's initial input. At this point, Sofia could have chosen to use the draft produced by ChatGPT directly, but this is not what she wanted. In the draft, there were arguments she did not understand, and others she did not agree with. Additionally, the tone was still far from matching exactly her style. This is why she then had multiple rounds of interactions with ChatGPT. Some of her responses are included below:

- I don't understand this part: "Moreover, legal obligations can mandate disclosure, such as in cases of data breaches where customer information is compromised. Ethical considerations, too, play a role in determining the threshold for public disclosure; moral imperatives often dictate that the public's right to know outweighs corporate secrecy."
- I don't think legal and ethical considerations are at the same level of ideas like the ones I shared. What other ideas do you have?

The final one-page reflection that Sofia submitted had little to do with the draft provided by ChatGPT. However, the process of leveraging ChatGPT to come up with a structure, test out different ideas, and brainstorm arguments was very helpful to write an essay with her own ideas and style in much less time than if she had written it completely on her own.

We argue that the use of ChatGPT from Angela and Sofia described just above on balance helped them learn. ChatGPT acted as a companion to help them brainstorm topics and guided them to structure their ideas. However, we also acknowledge that you might argue that they would have been better off not using ChatGPT for some of the things that they used it for.

Supporting or performing data analysis and visualizations

One area that has been completely revolutionized by the advent of ChatGPT is the field of data analysis. In Chapter 5, we illustrated how you can use ChatGPT to analyze student responses from a survey (including visualizing their answers in a graph). This was a simple use that hopefully gave you a taste of what's possible. But a lot more is possible, and the capabilities of ChatGPT in this area are rapidly advancing. You can now upload data sets from your computer, Google Drive or Microsoft OneDrive and within minutes produce statistical analyses and visualizations without writing a single formula or line of code. This field is both niche and rich in scope, exceeding what we can cover in this book. It is however very impressive, and the companion site has links to resources in this area if you are interested in exploring.

We want to offer some caveats. First, if the user (you or the student) does not have experience with statistical analysis and/or coding, it might be hard to verify whether the process or output that ChatGPT generated is technically accurate. As always, verifying ChatGPT's accuracy is important and this requires some skills. Second, if one of your course goals is to teach students how to code to produce statistical analyses, ChatGPT might detract students from achieving this goal as ChatGPT could be giving them the answers without teaching them the skills to get to these answers. Third, the usual caveat of data analysis applies: "Garbage in, garbage out." If the data you upload into ChatGPT is of low quality (e.g. poorly structured, full of errors...), the output, no matter how quickly and nicely is produced, will not be of high quality.

These caveats aside, ChatGPT can offer immense value to make the process of analyzing data (whether in Excel or a coding language like R or Python), more efficient for your students. Here are a few examples of how ChatGPT can help in the process of learning to analyze data:

- Carry out tedious data processes so students can focus on the conceptual and analytical understanding of the data analysis process
- Suggest ways to clean or structure a dataset and implement them
- Visualize data into graphs
- Explain how an existing data set is structured
- Explain a data visualization such as a graph, map...
- Debug coding errors

Summarizing a text passage

Students are often asked to read papers before a class and many instructors complain that they rarely do. We hypothesize this has likely happened to you too. Conversely, many students complain that such pre-class readings are too long, that they don't know what portions to focus on, and that they don't have sufficient time to read everything all their instructors assign. The reality is that reading papers effectively requires skills that educators rarely explicitly teach to their students: to be able to skim the paper for the main ideas, to extract the main takeaways, to analyze the arguments and counterarguments critically, and to assess what sets the paper apart.

Summarizing papers is one of the most common student uses of ChatGPT. Users can either upload a PDF, provide ChatGPT with the link to a website, or copy and paste the text into the chatbot. This is another area where the impact of ChatGPT on the learning process is unclear. If this acts as an incentive for students to substitute reading the full paper in detail, then simply asking ChatGPT to summarize it may be detrimental to their learning. However, if this provides an incentive to at least read the summary instead of not doing the reading at all, then this use may actually help to enhance their learning.

We discussed with Shoroqu Othman, undergraduate student at Georgia Institute of Technology, how she used ChatGPT to summarize some of her pre-class readings. For example, she had a class called "Introduction to Media Studies" which typically assigned various readings of more than 10 pages each for each class. Then the class was used to discuss the readings. With all the commitments and work that students have, you can imagine how doing these readings in depth for each class could be overwhelming. She shared: "The quickest way to understand the key points and sift through the documents was to have ChatGPT summarize them, and then, time permitting, go through the documents very briefly to pull noteworthy quotes."

Beyond summarizing the text, students can also use ChatGPT to ask questions about specific segments of the text, clarify terms, assess the robustness of the methodology, interpret a graph, etc. This use of ChatGPT as a "reading assistant" is very common as well, and is another example of a use case where it's debatable whether it is good for learning. If you are interested in learning more about this debate, we recommend listening to Derek Bruff's excellent podcast titled "Intentional Teaching" (linked in the companion site).

Taking notes

Note-taking is an important part of the student experience in most classes. ChatGPT can help in this process by transcribing full lessons if they are available online or students have a recording, organizing existing notes, and highlighting specific actions in the notes in a given format (e.g. "bold any action items").

If this is an area that you are interested in, we recommend Derek Bruff's podcast episode with Marc Watkins from University of Mississippi, where they discuss the potential learning gains and losses of automating note-taking. This episode is also linked in our companion site.

Automating tasks

ChatGPT can help students streamline repetitive tasks. Peter Huette, a student at Harvard Kennedy School, used it to streamline the process of compiling and producing a weekly news briefing that she prepared for Meghan O'Sullivan, professor at Harvard Kennedy School. Every week, he would do research across different papers to find the most relevant global geopolitical and energy events and summarize the key insights into an email that he would send to Meghan.

Peter shared he was using ChatGPT more frequently than Google these days, and that he had integrated the use of ChatGPT into many of his work processes (e.g. creating outlines, proofreading...). When he realized that he could automate the weekly news briefing, he took the opportunity to experiment with ChatGPT.

In his initial prompt, he described the role that he needed ChatGPT to play and provided examples of news briefings he had created manually in the past. His conclusion? "It worked really well!"

 Peter Huette (Harvard Kennedy School Student)

Task: Generate weekly summaries of geopolitical and energy-related events from scraped news articles.

Instructions: I will upload the articles that I want to be included in the newsletter

Analyze: For each article, extract the main event or development, and identify the involved parties and their roles.

Summarize: Produce a concise paragraph summarizing the main points of the article, maintaining an objective and analytical tone.

Assess relevance: Evaluate and explain the relevance of the event or development in the broader context of global energy and geopolitical trends.

Format Output: Organize the information into sections as seen in the "Weekly Energy Briefing" that have been uploaded to your database

Context: Brief background or developments leading to the event.

Relevance: Detailed analysis of the implications and significance of the event within the energy and geopolitical spheres.

As you can see in the prompt above, Peter provides step-by-step guidance on how to analyze the articles that will be included in the newsletter and the format that ChatGPT should use to produce the output.

Key takeaways

- ChatGPT can support student in the process of getting feedback, learning a new topic, concept or skill, deepening their reflections, finding arguments and counterarguments, practicing deliberately, analyzing data, and accomplishing tasks more efficiently.

- ChatGPT does not enhance or hinder learning itself - it depends on how ChatGPT is used. If well used, it can empower students, provide tailored assistance, and enable active learning simulations that would not be available to students without this technology.

- The use of ChatGPT will increasingly be integrated as one additional step in the process of learning or completing assignments. Knowing how to integrate the technology in a way that enhances learning- e.g. asking ChatGPT to adopt a persona to debate a specific topic with instead of asking for the arguments and counterarguments to a thesis - will be a skill in itself.

Chapter #9 - Nudging students to learn with ChatGPT

As an educator, you may have heard about how a colleague uses AI to provide one-on-one tutoring to their students, which might have sparked your curiosity about replicating their initiative. Alternatively, you might have also worried about the inequality that is emerging in your class, with some students lagging in their use of ChatGPT and missing out on valuable skills. Potentially, you may have read through Chapter 8 and be excited about the different ways in which your students can use ChatGPT to learn.

This chapter guides you through how you can encourage your students to use ChatGPT effectively to learn, building on the lessons from Chapter 8. Specifically, we focus on three topics. First, we share a framework to think about how to encourage your students to use ChatGPT and how to monitor their use. Second, we provide examples of educators who are nudging their students to use ChatGPT in ways that enhance their learning. Third, we present examples of educators who are purposefully helping their students build AI skills.

Building customized chatbots is one effective strategy that some educators use to encourage their students to use ChatGPT responsibly. Given the importance, complexity, and breadth of customized chatbots, we will explore this topic in depth in Chapter 10.

1. A Framework to Nudge Your Students to Learn with ChatGPT

If, as an educator, you cannot judge if a student has used ChatGPT, how could you even help them use it effectively for their learning? To address this issue, we have found it useful to think about the degree to which we can guide and monitor student use of ChatGPT as we plan classroom activities. A summary framework is included in Figure 9.1.

Figure 9.1 - Degrees of Guiding and Monitoring Student Use of ChatGPT

Degree of Guidance

Low   High

| Ask them to use ChatGPT | Describe how you expect them to use ChatGPT | Provide the specific prompt they should use | Create a customized chatbot for your students |

Degree of Monitoring

Low   High

| Ask for a reflection on the use of ChatGPT | Ask students to report how they used ChatGPT | Ask for and review link to ChatGPT conversation |

Degree of guidance: It helps to be clear and explicit about when and how students can use ChatGPT. As we discussed in the introduction to Part III, an inexistent policy may be interpreted as an invitation to use ChatGPT freely, which gives way to potential uses that are detrimental to learning. Beyond being clear that they can use ChatGPT (when appropriate), you can provide them with further guidance by describing how you expect them to use it, providing a specific prompt that they should use, or even creating a customized chatbot (see Chapter 10 for more on this). You will have an opportunity to see this in practice in the next section.

Degree of monitoring: It may be useful to design some form of assessment or monitoring of student use of ChatGPT. We have identified three main ways educators have approached this issue: asking students to reflect on their use of ChatGPT and how it has impacted their learning, asking them to report how they used ChatGPT, and asking them to share the link to their conversation with ChatGPT. These solutions foster opportunities for honest conversations and exchanges between you as an educator and your students as you all navigate this evolving landscape. It is important to note, however, that none of these alternatives can fully guarantee that students are using ChatGPT solely to enhance their learning. Students may still misrepresent their use of ChatGPT in their reports. Moreover, documenting AI use is becoming increasingly difficult in an era where it is being baked into the software we use every day (Word, Google Doc, etc.).

2. Encourage Students to Boost their Learning with ChatGPT

One way to integrate ChatGPT into your students' learning journey is by encouraging them to use it in any of the ways that were described in Chapter 8. As a reminder, these uses are the following:

1. Getting feedback
2. Learning a new topic, concept, or skill
3. Deepening reflections
4. Finding arguments and counterarguments
5. Practicing deliberately
6. Accomplishing tasks

With this in mind, we will now walk you through several examples of how different educators have encouraged their students to use ChatGPT for learning. While we will not cover examples for each of the six uses, we hope that these will be enough to inspire you to try something new in your teaching.

Encourage Students to Get Feedback from ChatGPT

Dan taught a course called "Thinking Analytically in an Uncertain World" in which he experimented with various ways to encourage students to use ChatGPT. In previous chapters, we shared how he encouraged his students to use it to learn about a specific topic. In a different assignment for this course, students had to answer a question that required applying a framework they had learned in class. A follow-up question asked them to improve their answer using ChatGPT: "Use generative AI (e.g. ChatGPT) to help you improve your answer to the previous question. Write how it improved your answer and please share your learning in our Slack #gen-ai channel." This allowed students to see the extent to which ChatGPT could help enrich or improve their work while allowing them to learn from each other about different ways of using AI for learning. By sharing their reflections in Slack, Dan was also able to learn how much students perceived they were learning from using AI in the course.

Pary Fassihi, senior lecturer at Boston University, also experimented with encouraging her students to get feedback from ChatGPT. While Dan's guidance was very broad (e.g. he did not provide any specific prompt), Pary shared with her students dozens of structured and well-crafted prompts to obtain feedback. The purpose of this exercise was to help students receive feedback on their papers akin to that provided by a peer. She therefore recommended students start by asking ChatGPT to impersonate a peer providing them with feedback. Below is the prompt she suggested her students use.

 Student of Pary Fassihi (Boston University)

I have written a critical review paper on **[your topic]**, and I'd like for you to take on the role of my peer and review the paper for me based on a set of criteria I will give you. I will give you these criteria one by one, and I would like you to give me feedback and offer suggestions for improvement. Thank you! The first criterion is: **[insert criterion]**

The prompt above needed to be completed with specific criteria to guide ChatGPT in the process of providing feedback. She then outlined 9 criteria that her students should seek feedback on, along with a prompt for each criterion (included in Table 9.1). She encouraged her students to think critically and not take the feedback at face value by explicitly asking them to judge ChatGPT's feedback: "Some of ChatGPT feedback may be useful, but some may not! Please avoid changing everything it asks you to change, and make sure your voice and YOU still remain very present throughout your paper."

Table 9.1 - Pary Fassihi's Prompts for Students to Receive Feedback

Criteria	Prompt
Claim Clarity and Argumentation	How clear is the claim in articulating the paper's stance on the 'post-plagiarism era' and its implications on academic integrity and authorship?
Critical Engagement with Sources	Does the paper critically engage with Sarah Eaton and Maha Bali's perspectives on academic integrity in the digital age? How can this analysis be improved?
Evidence and Support	Evaluate the evidence used to support the main argument. Is the evidence relevant, sufficient, and effectively integrated into the argument?
Analysis and Insight	How well does the paper analyze the implications of digital technology on academic integrity and authorship? What insights or unique perspectives does the paper offer?
Organization and Structure	Assess the organization of the paper. Is the argument presented in a logical, coherent manner that is easy to follow?
Originality and Creativity	Does the paper present an original viewpoint on the 'post-plagiarism era'? How does it contribute to the existing conversation on this topic?
Use of Language and Style:	Evaluate the clarity, precision, and appropriateness of the language and style. How can the readability of the paper be improved?

Conclusion and Implications	Is the conclusion effective in summarizing the main points and reinforcing the paper's stance? Does it discuss the broader implications of the findings?
Recommendations for Improvement	Based on your assessment, what are the top three recommendations you would make to improve this paper?

Source: Based on Pary Fassihi's Peer Review Papers assignment for course WR152 at Boston University

After obtaining feedback on all criteria, she asked her students to write a one-paragraph reflection on what they learned from the interaction, what they changed, what they decided to keep, and why.

The final example comes from Fernando Díaz del Castillo, Chief Learning Officer at Mentu, a digital learning ecosystem dedicated to closing the education gap in Latin America. Fernando experimented with asking his students in an independent school to use AI to receive feedback on the content and quality of their writing for a class project. He then asked students to share their interaction with the AI tool and to reflect on what parts of the exercise had been useful in helping them improve the quality of their work. His abridged instructions are included in Figure 9.2.

Figure 9.2 - Fernando Díaz del Castillo Encourages Students to Receive Feedback from a Gen AI Tool

In this task, you will use AI to review your work on the project. You will craft two prompts to help you review your individual product and you will post the results of that interaction, your prompt, and a final reflection on the use of AI as a result of the task.

1. Craft two prompts, using your best understanding of prompt engineering, to help you review your individual product. The first must focus on improving the depth of content

of your writing and the second should be focused on helping you improve the quality of the writing to make it engaging for the specific audience of your podcast.

2. Select an AI (ChatGPT, Claude, or Gemini) and use the prompts to review your individual work. Have a conversation with the AI.

Go to Knowledge Forum and in the view with your project plan post a new note with the title "AI Review". Copy and paste the whole interaction with the AI, including all prompts and responses.

"Build-on" that note (create a connected note) and answer the following questions:

    a) How have you used AI throughout your project?

    b) To what extent has it been helpful?

    c) What are two (2) things you learned about AI in this class?

Encourage Students to Learn a New Topic, Concept, or Skill with ChatGPT

You could ask your students to use ChatGPT, either individually or in small groups, to learn about particular topics that you would like to delve into later in class. In this way, each student (or group) can adjust the learning to their preferences. This is what Robert Klitgaard, a professor at Claremont Graduate University, did in his two courses in the Fall of 2023. Robert documented his experimentation in a fabulous paper he wrote in early 2024 titled "Using ChatGPT in Graduate Education: A Beginner's Guide (and We're All Beginners Here)."[16]

Robert used ChatGPT in his teaching in different ways, but we will focus on how he used it inside the classroom. Throughout the semester, he asked his students to learn about a topic in class using ChatGPT and provided them with a template prompt. After students had done this, he invited a classwide conversation about the topic and about how helpful ChatGPT was in the learning process. Below are some of the many prompts that Robert asked his students to use during class. **Bolded** segments were meant for students to customize, but more generally he encouraged his students to customize any part of the prompt. For a fuller set of prompts that Robert used, please check his paper (linked in our companion site).

Student of Robert Klitgaard (Claremont Graduate University)

I am a graduate student in **[your subject]**. Please explain the concept of externalities for me. Please provide two examples.

Follow-up:

How might cultural diversity create positive and negative externalities? Please give examples of each.

Student of Robert Klitgaard (Claremont Graduate University)

You are a kind, helpful statistician who will help me understand **[regression discontinuity design (RDD)]**. Please begin by introducing yourself. Then ask me about my level of familiarity with **[RDD]**. After I answer, tailor your explanation to my level of familiarity. Provide the theory and give an example. Ask me if I have questions, and then please continue helping me learn until I ask you to stop.

Student of Robert Klitgaard (Claremont Graduate University)

You are an expert in evaluation and policy analysis. I am **[describe your role and the challenge you face combining "outside" expertise and "local" knowledge]**. Please help me design a convening to combine outside expertise and local knowledge. The convening should include stakeholders such as my organization's experts, local community leaders, and local businesspeople. My goal is to bring together these stakeholders and help them problem-solve creatively. The convening consists of four stages:

Stage 1. Finding and presenting data that identify the local challenges and help participants compare their situation with other places around the world.

Stage 2. Finding a success story from elsewhere where outside knowledge and expertise was successfully combined with local knowledge and know-how, and then conveying the story in the part A, part B style of a Harvard Business School teaching case.

Stage 3. Creating and then conveying to participants a simple theory of change to help participants work through the options.

Stage 4. Considering the local context, creating and then discussing with participants an imaginary news story describing their success five years from now.

Please use these four stages in a conversation with me about designing a convening. Beginning with stage 1, please help me figure out what to do in each stage. Feel free to ask me questions. After we finish discussing stage 1, then let's move the conversation to stage 2. Pause after each of the four stages to ask me for comments or suggestions. Do you understand? Are you ready to begin?

By sharing prompt templates and allowing part of the class conversation to be about the use of AI, Robert was helping his students learn how to leverage AI in their learning, research, and life. Students were overwhelmingly enthusiastic and positive about this experience. Anuradha Dhanasekara, a PhD student in Psychology, reported: "Embarking on this journey with ChatGPT proved transformative for both my academic and professional endeavors. During classes, it felt like participating in a dynamic group experiment, exploring not only the course content but also the academic potential of ChatGPT. We started with class prompts, analyzing and discussing the varied outputs, which revealed the tool's diverse capabilities. Learning to refine our prompts enabled us to tailor ChatGPT's responses to our specific needs. Engaging with ChatGPT conversationally, whether as an expert, mentor, assistant, or coach, was both enjoyable and invaluable. I cherished having this versatile 'friend' —who is always available, intelligent, and never overwhelmed— to brainstorm ideas and discuss topics. Additionally, ChatGPT's ability to succinctly summarize materials was a lifesaver during time-crunched reviews."

Chasen Jeffries, PhD student in International Political Economy and Computational Analytics, commented: "Integrating ChatGPT into our coursework proved to be as engaging as having a team of colleagues constantly available for discussions. I learned a great deal from Dr. Klitgaard's expert utilization of ChatGPT in our classes." Chasen then added "In our course, the use of ChatGPT was particularly beneficial due to its application in highly specific contexts. Professor Klitgaard often provided specific prompts that fostered intellectual discourse and allowed for the personalization of topics. For instance, I could delve into public-private partnerships in a field of interest engaging ChatGPT as a subject matter expert offering diverse viewpoints. This approach not only introduced me to new concepts but also illuminated various possibilities that would have taken hours of brainstorming or research on my own. Furthermore, the debates with ChatGPT encouraged me to challenge ideas, formulate counterarguments, and better comprehend their underlying logic. This interaction was invaluable in clarifying and strengthening my understanding of both my viewpoints and opposing perspectives."

Encourage Students to Deepen Their Reflections with ChatGPT

Mitchell Weiss, professor of Public Entrepreneurship at Harvard Business School, encouraged his students to deepen their end-of-class reflections using ChatGPT. His goal was to invite his students to reflect on what they had learned in the course, solidify concepts, and think about how they would apply what they learned in their work.

As Pary and Robert did, Mitchell also shared a ready-to-use prompt with his students. You can read it below.

Student of Mitchell Weiss (Harvard Business School)

I am a student in the Public Entrepreneurship course at Harvard Business School. The course finishes this week. I would like your help reflecting on the course what I have learned and what I think. I am going to list five questions here. I would like you to ask them to me, in sequence and one at a time. I would like you to wait for my answer to each. And then I would like you to ask me a follow-up question that would deepen my reflection. Please continue to wait for my answers and then ask me a follow-up question until I ask you to move on to the next numbered question I gave you here.

Don't move on to the next numbered question until I ask you to, please. And also, please don't provide any commentary on my responses besides your follow-up questions; I want this really to be an exercise where I reflect, and your probing questions will help me.

One: If you had to explain to a friend or relative what this course was about, what would you say?

Two: What is one thing from the course you hope to remember for the rest of your life?

Three: What is one topic or issue on which you changed your mind throughout the course?

Four: What is one riddle or question about which you still feel unsettled, where you still haven't made up your mind?

Five: What is one way in which you hope to pursue "public entrepreneurship" in your life as it unfolds?

Please note how the added value of doing this reflection with ChatGPT vs. in a Google Form or a simple document is that ChatGPT can ask follow-up questions that deepen students' reflections. Mitchell therefore made sure to include this instruction as part of the prompt above. To give you a sense of what the experience as a student might look like, below is a short description.

ChatGPT

Let's start with the first question.

One: If you had to explain to a friend or relative what this course was about, what would you say?

 Student of Mitchell Weiss (Harvard Business School)

The course was about moving towards a "possibility government", a more innovative government, and beyond the status quo.

ChatGPT followed up with various questions to deepen the student's reflection:

- What aspects or examples of "possibility government" were most striking or inspiring to you during the course?
- Can you elaborate on a specific example of government innovation during COVID-19 that you found particularly impactful or surprising?
- Why do you think creativity in government tends to emerge more in times of crisis, and what might be done to encourage such innovation during more stable periods?

Once the student felt like the question had been answered in enough depth, they could simply ask ChatGPT to move to the next question. This ensured that the reflection exercise remained engaging and avoided becoming tiring or frustrating.

Mitchell had to decide whether to ask his students to share their conversations with ChatGPT (which they can do by sharing a link to the conversation or simply copying and pasting them into a document). He sees a tradeoff. On the one hand, doing so would have given him a good sense of what the students got out of them and perhaps created incentives for his students to exert effort in this activity. On the other hand, he wanted the reflection to be truly personal, where his students could be honest and not feel the need to perform in a certain way for their teacher. In the end, he opted not to request his students to share their reflections, but you could imagine him doing so particularly if he had requested these reflections in the middle of the semester and might have been able to follow up with individual or group conversations about them.

Encourage Students to Practice Deliberately with ChatGPT

As an educator, you could encourage your students to create new practice questions with ChatGPT so they practice and apply their knowledge. The most efficient way to do this is through a customized chatbot, a method that you can learn more from in Chapter 10.

Another interesting way to incorporate ChatGPT in your classroom to help students practice deliberately is through role-playing or simulations that encourage hands-on practice. Whether conversing with a historical character or designing an interactive learning experience by asking ChatGPT to adopt a specific persona, this use case captures the essence of the excitement of integrating AI into the classroom.

The first example features Mónica Flores Rojas, from Universidad Católica Boliviana. Mónica teaches English courses and wanted to empower her students to practice interviewing in English and be better prepared to face the cultural differences of interviewing for a job in the United States. As part of a series of workshops designed with this goal in mind, she asked her students to use a series of prompts in ChatGPT that would prepare them to have an interview in English.

 Student of Mónica Flores Rojas (Universidad Católica Boliviana)

Can you create a table that shows the differences and similarities between having a job interview in Bolivia and in the United States? Focus on cultural differences.

The next step was to ask ChatGPT to provide a list of common questions that appear in job interviews and to be prepared to answer them. This was followed by an activity consisting of using ChatGPT to simulate a job interview through a back-and-forth conversation.

 Student of Mónica Flores Rojas (Universidad Católica Boliviana de la Paz)

Act as a manager of a big company and I'll act as a job applicant. Wait for my answer before you ask the next question. I want to practice my job interview skills.

To wrap up this series of workshops, Mónica carried out short job interviews with each program participant and evaluated their readiness levels considering linguistic and cultural aspects. Using ChatGPT helped Mónica provide her students with information about complex aspects of the job interview process. As Mónica shared, this would have required longer for students to discern if done without the help of technology, and the results were very positive.

The second example comes again from Mitchell Weiss, professor at the Harvard Business School, who developed an ingenious exercise for his students in his public entrepreneurship class to examine the extent to which generative AI could be useful in solving public problems. The exercise, named "Storrowed" and now published as a Harvard Business School case (see companion site), examines a problem in Boston when drivers of overheight trucks get their trucks wedged under the bridges on a local highway called Storrow Drive (hence the name "Storrowed"). Mitchell opened the class by showing the students images of a news story of a truck being storrowed.

Mitchell then asked students to use AI (in his case a private sandbox connected to ChatGPT and other LLMs) during class (in small groups) to do two things: understand the nature of the problem and look for solutions. He split the room in half. One side took the role of government officials, and the other side took the role of tech entrepreneurs. Part of the goal of the class was to see how each side would frame the problem and be attracted to different solutions. After giving students some time to prompt the AI tool in small groups, Mitchell led a discussion in which he asked a series of substantive questions (e.g., What do we think is the problem here?) and students gave their substantive answers along with a description of the prompt(s) they gave to the AI tool that informed that answer. Thus, Mitchell helped his students not only develop an understanding of how to tackle this public problem but also how to use generative AI to help them tackle this problem.

There is an interesting backdrop to this story when Mitchell taught this class for the first time. He woke up the morning of class wanting to raise the stakes. He asked himself: "How do I create some tension, some energy, some urgency, some expectation?" He created a game. Without any coding experience, he asked ChatGPT to write in HTML a game where the students would see two trucks that were heading towards a bridge and would hit the bridge in 30 minutes unless Mitchell clicked a button that would send the trucks back 10 seconds (see visual below in Figure 9.3). An updated and customizable version of the game is available for instructors as part of the Harvard Business School case.

Figure 9.3 - Game Created on ChatGPT by Mitchell Weiss to Raise the Stakes in His Classroom

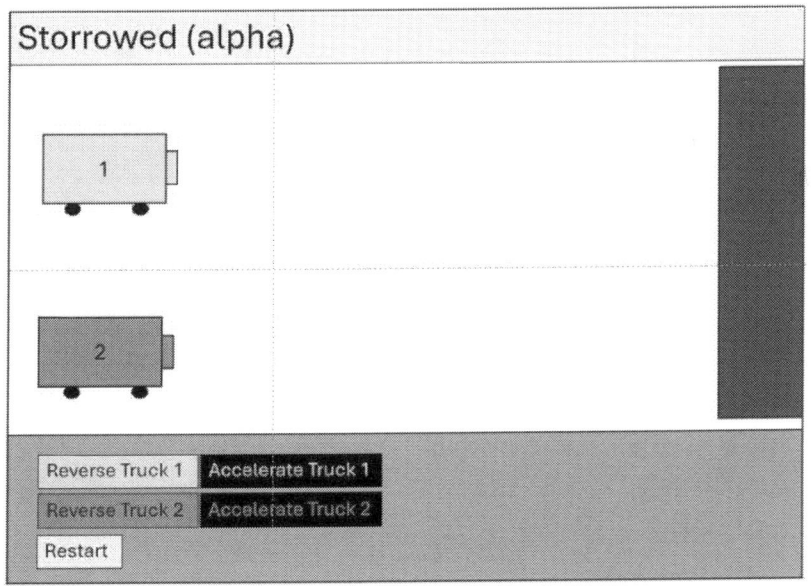

The pedagogic goal of this game was to encourage students to use good prompts. He raised the stakes by setting this as a contest between government officials and tech entrepreneurs: the first one to get storrowed would lose. The game worked as follows. One of the trucks represented the government officials and the other truck represented the tech entrepreneurs. When a government official made a good prompt (e.g., Why does storrowing happen in Boston?), Mitchell would move their truck back 10 seconds (thus delaying the time when they would get storrowed). If they made a very good prompt (e.g., Can you do a 5 Whys analysis on why storrowing happens in Boston?), Mitchell would move their truck back 30 seconds. The greater the number of good prompts made by your team, the more your truck got delayed to reach the overpass. Conversely, when a student was a victim of AI hallucination, Mitchell would move their truck forward 10 seconds. Students were able to see how tech entrepreneurs and government officials defined the problem differently and proposed different solutions.

Mitchell has now taught this case many times and describes two lessons that he learned in the process:

1. Before anyone weighs in on how these AI tools should and shouldn't be used, they need to know how they could and couldn't be used. Proficiency is a first step.
2. One of the sources of hesitance for people in using the tools (like fear, morality, etc.) is a worry about self-efficacy. The exercise helps show people they are capable. That turns out to be important as a pathway towards more use and exploration.

You might not be willing to develop an exercise as elaborate as Mitchell's but you might want to think more broadly how you can help your students learn how to leverage AI tools to develop the skill you are trying to teach them. This is especially valuable if you envision this use as a realistic representation of how your students will exercise the skill after they graduate.

The third and final example is based on a game designed by Matthew Wemyss, Assistant School Director at Cambridge School of Bucharest. Matthew designed a prompt that allows students to learn about historical events through a simulation game. To illustrate the power of Matthew's game, let's imagine that a hypothetical student of Matthew is taking a course on the History of International Relations. As part of the course, he needs to learn about the Bretton Woods Conference in 1944, which led to the creation of the IMF and the organization that later became the World Bank. This conference established the norms that would rule the international commercial and financial relations of the post-war era.

In this example, Matthew created the prompt that gives ChatGPT the instructions to get the game started. As always, this prompt is linked in the companion site. In Chapter 10, you will learn how this could have been done through a customized chatbot to improve the student experience.

 Student of Matthew Wemyss (Cambridge School of Bucharest)

You are an expert in designing text-based adventure games for educational purposes. Please follow these detailed steps.

Ask me to provide the following information:

Age of the students

Subject being taught

Learning Objectives

Refrain from generating the game until I have given you the required details.

After I have provided the Year Group, Subject, and Learning Objectives, generate a text-based adventure game based upon the subject and learning objectives that follows the sets of rules below:

Presentation Rules:

The game output will always show 'Location', 'Description', and 'Possible Commands'.

Always wait for the player's next command.

Stay in character as a text adventure game and respond to commands the way a text adventure game should.

The 'Description' must stay between 3 to 10 sentences. The 'Description' should include educational information appropriate for the year group of the player.'Possible Commands' should have letter assignments that the player can press to select the command they wish to action (for example a.[command], b.[command])

Fundamental Game Mechanics/rules:

1. The game presents a story with branching paths. Players make choices at key decision points, which lead to different outcomes and directions in the story.

2. Players are presented with options at various points in the story. Each choice affects the subsequent events and outcomes, leading to different story arcs or endings.

3. Players assume the role of a character within the story. Their choices reflect the character's decisions and actions, shaping the course of the narrative.

4. The game can have multiple possible endings, depending on the choices made by the player. These endings can range from successful outcomes to more challenging or surprising conclusions.

5. The game's narrative is presented through descriptive and engaging text. The language used in the adventure must be immersive and help them make informed decisions, but be appropriate for the age of the player indicated by the Year group they are in.

6. At a key part of the adventure, players encounter multiple-choice questions related to the learning objectives. At these points, they must answer the questions before they can take any other action. If they answer correctly, they can progress; if not, you will reteach them the knowledge they need to know, followed by another question until they answer correctly. At these points, the 'Possible Commands' presentation rule is suspended and the user must type in their answer to the question. Once the answer is correct, and they

progress the 'Possible Commands' rule is enabled again.

7. The game provides feedback on the consequences of players' choices, including immediate feedback on their answers to educational questions. Correct answers lead to progression, and incorrect answers trigger hints and retries.

8. The player can only exit the game when they have completed the adventure.

Rules of the setting:

Populate the game world with interactive characters, marking their dialogues with quotation marks.
Craft suitable names for each character.
Refer back to these rules after every prompt.
Start Game.

The instructions specify that ChatGPT should always gather three critical pieces of information to design the game: the age of the students, the subject being taught, and the learning objectives. These pieces of information could have also been provided as part of the prompt if the professor had desired to do so. Below is ChatGPT's response.

 ChatGPT

To start designing a customized text-based adventure game that meets your educational goals, I need the following information:

1. **Year Group of the Students:** Please specify the age or grade level of the students who will be playing this game.

2. **Subject Being Taught:** Let me know the academic subject or specific topic the game should focus on.

3. **Learning Objectives:** Describe the key knowledge or skills you want the students to

develop through playing this game.

Please provide these details so I can create a game that is both engaging and educational.

 Student of Matthew Wemyss (Cambridge School of Bucharest)

1) 20 years old 2) History of International Relations (3) Learn about the creation of the Bretton Woods system.

Once the key details have been provided, the game has started. It is designed to be fully interactive, so ChatGPT continuously provides the students with input so they can choose their own adventure, and explore in practice what it may have been like to be part of the Bretton Woods Conference.

 ChatGPT

Thank you for providing the details for the text-based adventure game. Here's how I will design it:

- **Setting the Scene and Location:** The game will take place in various historic locations relevant to the Bretton Woods Conference of 1944. Players will assume the role of a young diplomat navigating through crucial negotiations.
- **Developing the Narrative:** The narrative will engage the player in the complex political atmosphere of post-World War II, where they must use their knowledge of international relations to influence the outcomes of the Bretton Woods system.
- **Implementing Learning Objectives:** Throughout the game, players will encounter multiple-choice questions and scenarios that require understanding the motives, principles, and impacts of the Bretton Woods system. Correct answers will advance the story, while incorrect answers will provide educational feedback and additional chances to answer correctly.

- **Character Interaction:** Players will interact with historical figures involved in the conference. Each character will have distinct personalities and insights into the formation of the system.
- **Game Mechanics:** As per the rules, the game will present options at key decision points, involve educational questions at critical moments, and provide feedback based on player choices.
- **Ending:** Multiple endings will depend on the diplomatic paths chosen by the player, reflecting the real-world complexities of international agreements.

Now, let's begin the adventure!

Location: Hotel Lobby, Bretton Woods, New Hampshire, July 1944

Description: You step into the bustling lobby of the grand hotel, transformed into a hub of wartime diplomacy. Delegates from over forty nations gather, debating the future of international financial systems. Your mission is to navigate through these pivotal discussions, ensuring the establishment of a system that could prevent future economic crises. Historical figures like John Maynard Keynes and Harry Dexter White can be seen discussing intensely by the fireplace. Your choices here could influence the creation of the world's new economic order.

Possible Commands:

a. Approach John Maynard Keynes
b. Listen to Harry Dexter White's conversation
c. Review your notes on the Bretton Woods proposals
d. Observe the interactions quietly

What will you do next?

The student could now choose either of the four scenarios depending on their interests. Let's see what 'option a' (a conversation with John Maynard Keynes) would look like in this case.

Student of Matthew Wemyss (Cambridge School of Bucharest)

a

ChatGPT

Location: Conversing with John Maynard Keynes, Hotel Lounge

Description: As you approach, John Maynard Keynes, a towering figure in economics, looks up from his papers. "Ah, a fresh perspective! Tell me, what do you think is the fundamental issue we must address with this new system?" he asks, tapping his pen thoughtfully. You know this is your chance to demonstrate understanding and influence the proceedings. Around you, delegates argue over fixed exchange rates and the role of what will become the International Monetary Fund.

Possible Commands:

    a.   Discuss the importance of fixed exchange rates
    b.   Advocate for flexible exchange rates
    c.   Suggest focusing on economic stability post-war
    d.   Admit you need more information

What's your choice?

Student of Matthew Wemyss (Cambridge School of Bucharest)

b

274

ChatGPT

Location: Deep in Discussion with John Maynard Keynes, Hotel Lounge

Description: "Flexible exchange rates, you say?" Keynes raises an eyebrow, intrigued. "An unconventional choice given the current sentiment. Could you elaborate on why you believe flexibility might better support global economic recovery?" This is a critical moment to showcase your grasp of international economics and the stakes involved in setting up a system that can adapt to post-war realities. The room's atmosphere is tense, with every delegate keen to shape a resilient financial structure.

Educational Moment: Before you can respond to Keynes, it's essential to ensure you understand the concepts at play. Here's your question:

"What is a potential benefit of having flexible exchange rates in the international financial system?"

Type your answer: (Please type in your answer, and we'll continue based on your understanding. If incorrect, I'll provide further information and another chance to answer.)

Note that at this point, the game becomes an opportunity for the student to reflect on the pros and cons of one of the main points of discussion of the conference: whether currency exchanges should be pegged or let float. Furthermore, the description of the conversation creates a very vivid image of what it must have felt like to have these discussions during the conference. Let's imagine that the student has some ideas of how to answer this question based on what they remember from class.

 Student of Matthew Wemyss (Cambridge School of Bucharest)

Flexible exchange rates provide more flexibility for central banks to set their own monetary

policy

 ChatGPT

Feedback: Correct! Flexible exchange rates do indeed provide more autonomy for central banks to implement monetary policies tailored to their country's economic needs without being tied to maintaining a fixed exchange rate. This flexibility can be crucial during economic shocks.

Continuation of the Discussion with Keynes:

John Maynard Keynes nods appreciatively, understanding the depth of your insight. "Indeed, that autonomy can be crucial in times of economic distress. It allows countries to respond more effectively to their unique situations without the constraints of maintaining a fixed rate. However, it also presents challenges, such as potential currency volatility. Let's delve deeper. How might this impact international investments?"

Possible Commands:

a. Argue that volatility can deter investment, affecting global economic stability.

b. Suggest that while volatility is a risk, the potential for higher returns might attract more daring investors.

c. Discuss mechanisms that could mitigate these risks without sacrificing monetary policy flexibility.

d. Request more time to gather data on historical impacts.

What will you do next?

The conversation can continue for as long as the student desires, taking them to explore Keynes's own opinion on the exchange rate regime, approach other delegates at the conference to inquire about their concerns, or make their own suggestions about the regime that should be set forth and gather feedback from various historical characters. Throughout this exploration, the student has the agency to choose what to delve further into, what to explore, and how to interact with characters in this historical event.

Encourage Students to Accomplish Tasks with ChatGPT

The final way in which students can use ChatGPT is simply as a tool to accomplish tasks more efficiently. Some educators encourage this type of use and nudge students to reflect on how this use is affecting their learning and knowledge acquisition process.

For example, David Dockterman, lecturer at the Harvard Graduate School of Education, teaches a course called "Impact by Design." In this course, students are asked to design projects on areas in need of innovative impact solutions. This activity involves six different steps with six corresponding deliverables:

1. Fall in love with a problem.
2. Unpack the impact ecosystem.
3. Target what change(s) you can address and for whom.
4. Generate ideas and a Theory of Action.
5. Create and test the most MINIMAL viable products of your theory of action.
6. Convince us you can innovate for impact.

One of the challenges David faced was how to get students as quickly as possible into the design work, the skills of iterating, questioning, and defining. To do that, students need to have a lot of background knowledge and understand a lot of content. David thought generative AI could help accelerate students through that problem definition and content acquisition phase so they could get more quickly into the work. David encouraged students to use ChatGPT by role modeling with specific examples of how ChatGPT may help them in each step of the process.

David concluded: "I have learned a lot since last summer when I first put together a plan for Gen AI in my Impact by Design course. For this coming academic year, I will continue to promote the use of Gen AI tools, including new features, like image and media generation, that might expedite rapid prototype development. Gen AI will not replace the requirement of reviewing primary research literature, talking with experts, or engaging directly with members of the impact ecosystem. I may, though, add a specific deliverable related to Gen AI use that ensures that all students get exposure to the tools."

3. Help Your Students Develop AI Literacy

As an educator, you may consider it important to build AI literacy in your students to better prepare them for the complexities and demands of the modern world. In an era where AI is increasingly permeating every aspect of our day-to-day lives, understanding its principles and applications can provide students with important professional skills.

Two educators that we interviewed for this book were thinking along these lines and designed various activities in their classrooms to help their students learn how to use ChatGPT. Below are their examples.

Fernando Díaz del Castillo teaches the course "World Religions" in a high school in Colombia and observed two challenges related to AI in his classroom. First, he realized that students were not aware of the risks involved in using AI, such as biases and hallucinations. Second, he observed that students using ChatGPT to complete activities for them were, in terms of Erick Klopfer and his colleagues at MIT, "bypassing important cognition" processes for learning.[17] To address these challenges, Fernando asked his students to do a basic learning exercise leveraging AI. In this assignment, they had to compare the answers they found to a set of simple questions such as "When and where did this religion originate?", "What is the current number of followers as a percentage of the world population?" or "What are the sacred/primary texts associated with this religion?". They did so using three different tools: ChatGPT, Gemini, and Google Search. Once they compared the answers, Fernando asked them to reflect on their use of AI. The responses were mixed in terms of the relative effectiveness of generative AI (i.e., ChatGPT and Gemini). What students did gain was a better awareness of how they can or can't use ChatGPT while still keeping the agency and effort on the learners' side.

Our second example comes from Lance Eaton, an instructional designer, and educational and social media consultant based in Providence, Rhode Island. He wanted to get his students to explore and assess how AI tools could be useful or not in his course. To do so, he encouraged students to explore various AI tools differently every week and created a document (available on our companion site) in which students could record their conversations with generative AI tools, along with reflections on what worked and what did not. For example, on week 1, students had to:

- Ask 10 completely different and unrelated questions within domains of knowledge that they are familiar with.
- Evaluate the quality of the answers and rate them on a 10-point scale (1 being completely inaccurate and/or useless; 10 being the most perfect and thorough answer).
- Review similarities in style and form across answers.

On week 2, students would transition to experimenting with different prompt styles with the following guidance:

- Ask the same question 10 different ways.
- With each question, consider changing the level of specificity, the style, etc.
- Observe similarities, differences, and surprises across the answers.

In the following weeks, students explored other uses of ChatGPT and continued reflecting on what was useful and what was not. In this way, Lance enabled his students to gain skills in both the subject of his course and the benefits and limitations of using generative AI.

✓ Key takeaways

- **Nudging students to learn with ChatGPT:** It is possible to encourage your students to use ChatGPT to enhance their learning. When designing activities that would involve the use of ChatGPT, it is useful to consider the degree of guidance you will provide (e.g. will you provide a specific prompt for them to use?) and the degree of monitoring you will implement (e.g. will you ask them to report how they used ChatGPT?).

- **How to nudge them:** You can encourage students to use ChatGPT in any of the six ways that we have explored so far: (1) Getting feedback (2) Learning a new topic, concept, or skill (3) Deepening reflections (4) Finding arguments and counterarguments (5) Practicing deliberately and (6) Accomplishing tasks.

- **AI Literacy:** Building AI literacy, like critical thinking or problem-solving, can be a learning goal worthy of its own to better prepare students for the complexities and demands of the modern world. Some educators encourage their students to experiment intentionally with different generative AI tools and learn key skills like prompt engineering.

Chapter #10 - Building Customized Chatbots

In this chapter, we delve into the world of building customized chatbots, a powerful feature to tailor ChatGPT to your specific instructional needs. If you are just being introduced to ChatGPT through this book, you might think this chapter is only for advanced users and be tempted to skip it. We encourage you to give it a try. We have sifted through the jargon and complexity to provide a straightforward explanation of how bots are built, what you can use them for, and examples of how other educators are creating their own. Bots represent the culmination of ChatGPT's potential in education, enabling educators to create personalized interactions that align with their curriculum and specific instructions, resulting in a more effective and relevant learning experience for their students.

1. An Introduction to Chatbots

Customizing your ChatGPT experience through custom instructions

So far in the book, we have used ChatGPT by submitting new prompts in new chats. This means that you need to give ChatGPT instructions from scratch. It works well because generally each chat is meant to help you accomplish a different goal. However, there may be instructions that you always want ChatGPT to follow regardless of the task. For example, you may find ChatGPT's style of responding too long-winded, and would like it to respond more concisely. Or maybe you would like ChatGPT to know that you are an educator so it can adjust its responses and tone to your background. Perhaps you find a particular ChatGPT habit irritating (e.g., its tendency to apologize) and want to eliminate it from its responses.

If you are interested in customizing your ChatGPT experience for all the chats you have with ChatGPT, you can accomplish this by creating custom instructions. You first click on your name (top right of the interface), select "Customize ChatGPT" and then answer these two questions:

- What would you like ChatGPT to know about you to provide better responses?
- How would you like ChatGPT to respond?

If you plan to use custom instructions, keep in mind that your responses to these two questions will apply to *all* your chats. We do not recommend being very specific because instructions that might be good for some chats might not be good for others. For example, you may use ChatGPT in both for personal and professional life. If you included your job description in your custom instructions, ChatGPT would still take this into account even if not relevant, such as when responding to personal queries. Neither of us uses custom instructions but if you are interested in using them, we suggest you experiment and explore whether the output you get with them is more to your liking. We have linked a couple of resources about custom instructions in the companion site in case you would like to explore further.

Customizing your ChatGPT experience through a customized chatbot

Sometimes, you might want ChatGPT to follow specific instructions repeatedly across several chats, but not in every chat. For instance, if you want ChatGPT to provide feedback on your class slides throughout the course, you would typically start a new chat for each session, repeatedly explaining the type of feedback you seek and uploading key documents like your syllabus. This process is time-consuming and inefficient.

When you need a set of instructions to apply to several, but not all, of your ChatGPT conversations, you can create a customized chatbot. This allows you to provide specific instructions and access data/documents that will be used each time you use the customized chatbot. Additionally, you can share customized chatbots with others for their use.

Customized chatbots created with ChatGPT are called Custom GPTs or simply GPTs. Custom GPTs are easy to build because no coding is required. This functionality of building Custom GPTs is reserved for paid users. However, users without a paid subscription can now use Custom GPTs that others have created. This means that if you have the paid ChatGPT version, you can create Custom GPTs for your students to use even if they don't have a paid account (more on this in section 3 below).

2. Building a Customized Chatbot for You

The best way to understand how customized chatbots work and how they can be helpful to you is to build one. Hence, we suggest that you create this chatbot with us following the instructions below. The goal of this chatbot is to help you create quizzes for your students. Even if this is not a use you find particularly appealing, we suggest you go ahead and create it so that you develop a feel for what is possible.

Step 1 - Create your Custom GPT

Begin in the main ChatGPT interface and click on "Explore GPTs" and then on the button in the top right corner that says "Create".

Figure 10.1 - How to Start Creating a Custom GPT

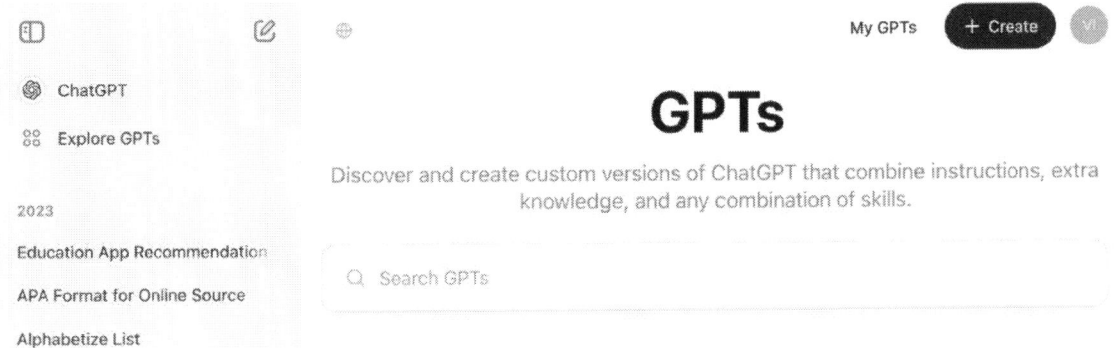

Step 2 - Configure your Custom GPT

There are two alternative ways to approach the configuration process. The first involves having a conversation with ChatGPT (by clicking on the "create" tab that appears in the screen on the left of Figure 10.2). The second involves filling in each instruction in the step-by-step process outlined in the tab "Configure" (i.e. right next to the "Create" tab on Figure 10.2). We recommend that you follow this "Configure" process as it will ensure that you consider all the key elements that help design your GPT.

Figure 10.2 - How to Configure Your Custom GPT

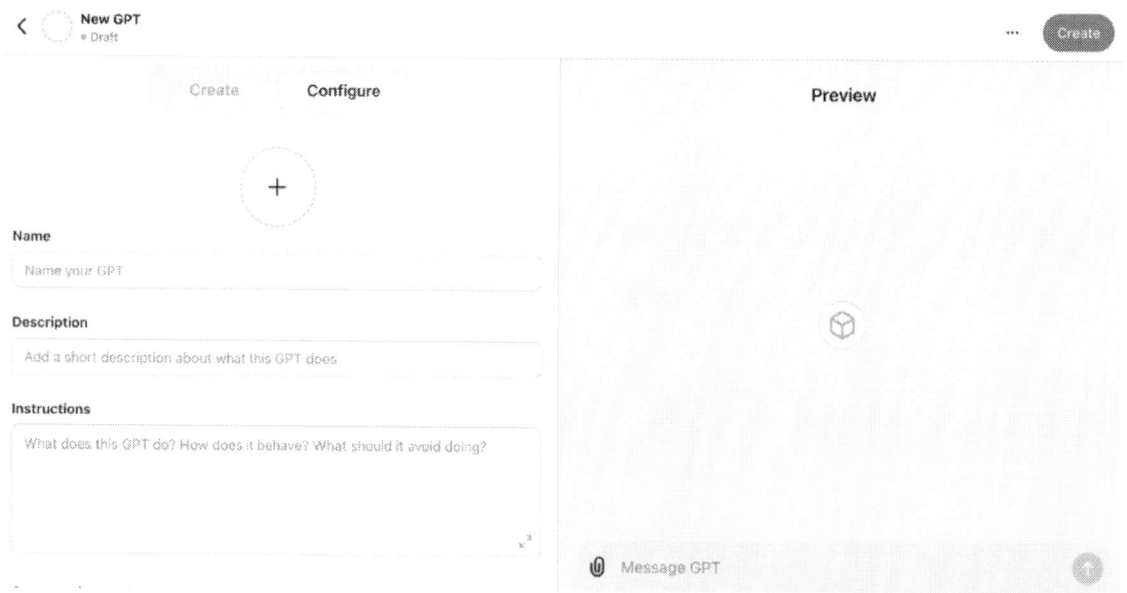

The step-by-step process will ask the following questions. We encourage you to use the answers listed in the second column of Table 10.1.

Table 10.1 - Step-by-step Configuration of Your Custom GPT

Question	Answer / Action
Name	My Quiz Assistant
Description	Create diagnostic quizzes for my classes
Instructions	*[Please copy and paste text from the box below; the text is also in our companion site]*
Conversation starters	*[Leave this one blank for now; given the goal of this Custom GPT, we don't really need it]*
Knowledge	*[Please upload your syllabus and other course material you consider relevant]*
Capabilities	*[Check all that are available]*

Below is the text you should include as part of the "Instructions" box. They are adapted from the Quiz Creator prompt from Mollick and Mollick (linked in the companion site); please fill in the bolded terms with relevant information for you and feel free to tweak it!

Instructions for Quiz Creator Custom GPT

You will act as my helpful teaching assistant and an expert in assessment. You create diagnostic quizzes that consist of multiple-choice and open-ended questions that test student knowledge. You will ask one question at a time and keep your part of the conversation brief.

Please take into account this information about me when designing your quizzes:

- I teach students at [**specify name or type of institution**]
- My students are [**describe your students briefly, i.e. high school, college, graduate, etc.**]
- The course I teach is [**specify name of the course(s)**]. I have attached the course syllabus

To begin with, please introduce yourself and ask me the following 3 questions:

Question 1 – What will the quiz be about?

Question 2 – What specifically (in 2-3 points) students should understand about this specific topic and what sticking points or difficulties students might have?

Question 3 – Would you like to attach any class material (slides, handouts, etc.) that I could use when preparing the quiz? [If I say that I want to attach material, wait until I attach it before proceeding].

Please wait for me to respond to these 3 questions. Do not move on until I respond. Do not ask any other questions until I respond. Do not mention topics or documents until I respond to the 3 questions.

Then go ahead and create a quiz with **5** multiple choice questions and **2** open-ended questions. The questions should be arranged from easiest to most difficult. Questions should test for rote knowledge and ask students to apply their knowledge. Do not focus on sticking points only. Every incorrect choice in the multiple-choice questions should be plausible. Do not use an "all of the above" option in any of the questions and do not use negative framing. When applicable, open-ended questions should prompt students to apply their knowledge and explain concepts in their own words and should include a metacognitive element, e.g., explain why you think this? What assumptions are you making?

Make the test nicely formatted for the students. Also give me an answer key. Explain your reasoning for each question and let me know that this is a draft and that you are happy to work with me to refine the questions.

Step 3 - Publish your Custom GPT

Once you have answered the questions to configure your GPT, you have three options to publish it: (i) make it available only for you, (ii) make it available to anyone with the link, and (iii) add it to the GPT Store so anyone can use it.

Given that this particular custom GPT is only for you, we suggest selecting the "Only Me" option.

Figure 10.3 - How to Share Your Custom GPT

Step 4 - Test your Custom GPT

Once your GPT is published, it is important to test it to ensure that it does what you want. In this case, try out building a quiz for one of your courses. If something does not work as expected, tweak the GPT's instructions by clicking on "Update" and test it again. The process of creating a GPT is meant to be iterative. As Teddy Svoronos, senior lecturer at the Harvard Kennedy School and designer of several excellent teaching chatbots puts it: "Test, iterate, test, iterate... The unpredictability of LLMs is what makes them so interesting, but also challenging to use in teaching. Have your teaching team, students you trust, or friends put each bot through its paces to see how it performs, where it goes wrong, and whether it can be "jailbroken" into doing things other than its intended purpose. Continually refine your prompt based on the feedback you receive."

Also, you can come back to edit your GPT at any time. To do so, click on "Explore GPTs" on the left pane, click on "My GPTs" and then click on the pen icon of the Custom GPT that you want to edit. Once you are back in the editing interface, you can go back to the "Configure" tab to adjust the instructions or go to the "Create" tab and let ChatGPT know what you don't like (e.g., the bot's answers are too long) and let it adjust the instructions. Either way the goal is to make changes and test again until you are satisfied.

Now that you have built your own custom GPT for you, we suggest you think about other possible Custom GPTs you could create to help you gain time or automate tasks you do repeatedly. Here are some ideas of when a customized bot might come in handy:

- You are using ChatGPT to provide feedback on your assignments, and find it repetitive to write a new prompt in a new chat for every assignment you want to review.
- You write many letters of recommendation that have roughly the same format and want to automate the tedious part and focus instead on the more authentic parts of writing them.
- You continuously provide the same feedback to your students and want help

automating the process (see Chapter 7 for more on the benefits and challenges of using ChatGPT for feedback assistance).

3. Building a Customized Chatbot for Your Students

You can also create customized chatbots for your students to use. The advantage of doing so is that you can tailor their responses to fit your instructional needs. This is therefore a powerful way to nudge your students to use ChatGPT with "guardrails" to enhance their learning. The process is the same as above, except that in step #3 you should share your GPT to anyone with a link. If you want to make your GPT available to anyone in the world, you should instead publish it in the GPT store. Additionally, OpenAI recently announced "ChatGPT Edu" which they positioned as a way for educational institutions to give you access to a version of ChatGPT that allows you to build Custom GPTs and share them only within your institution (ensuring greater privacy).

What might you create customized bots for your students for, you might ask? Tim Lindgren and his colleagues at Boston College have experimented with the creation of more than 10 different bots in their classrooms. After this experimentation, they provide a simple breakdown of four potential roles that a Custom GPT could play in the learning process.[18]

Figure 10.4 - Four Purposeful Roles for Course Chatbots

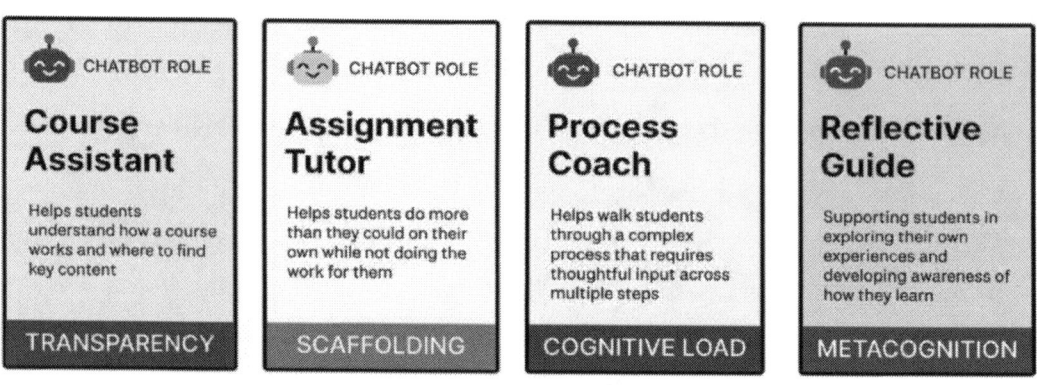

Source: Boston College, 2024

Each role is described below. These roles serve as an illustration, and we hope that this will encourage you to think creatively about how this technology can be helpful to you to teach more effectively and enhance your students' learning.

- **Course assistant:** helps students navigate the logistical aspects of a course so they can spend more effort on learning, such as how to sign up for office hours or which class slides explain a specific course concept.
- **Assignment tutor:** helps students by acting as a tutor that scaffolds learning and challenges students in their learning process.
- **Process coach:** provides step-by-step guidance to students during a complex process.
- **Reflective guide:** helps students scaffold the reflection process by providing structured guidance and encouraging them to deepen their reflection.

As you think about building your own custom GPTs for your students, you should be aware of three potential limitations. First, if your custom GPT is accessible via a link, anyone with the link can access it, even if they are not your students. This may or not be a concern depending on your circumstances. Second, custom GPTs are built on ChatGPT's technology and therefore subject to the same privacy concerns we outlined at the beginning of the book. We therefore recommend that no sensitive information is included in any chat with the GPTs. Third, as the creator of the custom GPT, you will not be able to access the conversations that students have with your custom GPT. This may be a problem, for example, if you want to understand how your students are using the custom GPT or if you are considering using it to record students' answers to exercises, such as reflections. If you are interested in ways of addressing these limitations, please explore section 5 below.

Illustrating the Differences Between ChatGPT and a Customized Chatbot

We now illustrate the power of customized chatbots for your students with a simple example. Imagine that a student named Blake is preparing for the SAT and wants to practice some math problems. Blake could interact directly with ChatGPT or through the customized chatbot built by Khan Academy called "Tutor Me." Blake submitted the same prompt to both.

Blake

I am preparing for the SAT. Help me solve this question:

Maria is staying at a hotel that charges $99.95 per night plus tax for a room. A tax of 8% is applied to the room rate, and an additional one-time untaxed fee of $5.00 is charged by the hotel. How would you represent Maria's total charge, in dollars, for staying x nights?

Below are their respective responses. Let's start by examining the response from ChatGPT (the general use tool we have been exploring throughout the book). While we generally encourage you to skim through ChatGPT's responses, this time we recommend that you take your time to read and compare the two different responses to the same prompt.

 ChatGPT (general use)

To solve the problem, we need to account for the costs associated with Maria's stay at the hotel, including the nightly room rate, tax, and the one-time fee. Here's the step-by-step breakdown:

1. Nightly Room Rate with Tax:

- The hotel charges $99.95 per night.

- There is an 8% tax applied to this room rate.

- To find the total nightly cost including tax, we multiply $99.95 by 1.08 (since 8% tax means we are paying 108% of the room rate).

- This gives us: 99.95*1.08.

2. Calculate the Total Room Charge for x Nights:

- The cost per night including tax is 99.95*1.08.

- For x nights, we multiply this amount by x: (99.95*1.08)x.

3. Add the Onetime Fee:

- The hotel also charges a one-time untaxed fee of $5.00, which we need to add to the total cost for the nights stayed.

4. Combine the Total Costs:

- The total cost for staying x nights is the sum of the total room charge and the one-time fee.

Putting it all together, we get the following expression:

Total charge = (99.95*1.08x)+5.00

Let's simplify this expression:

Total charge=(107.946x)+5.00

So, the total charge, in dollars, for staying x nights is:

107.946x+5.00

The general use of ChatGPT directly provides the full response to the question, leaving Blake with little work to do to engage productively in learning to solve this problem.

Now let's see the answer Blake would get from the customized chatbot "Tutor Me," designed to be a Socratic tutor.

 "Tutor Me" (customized chatbot)

Let's break down the problem step-by-step.
First, we know the cost per night is $99.95. What is the tax amount for one night?

Blake obtains a completely different answer from the "Tutor Me" chatbot. Instead of the solution, Blake receives a question that helps him by scaffolding the process of solving the problem. Let's say Blake responds to the question but gets the answer wrong. Instead of calculating 8% of the room rate, he calculates 80%.

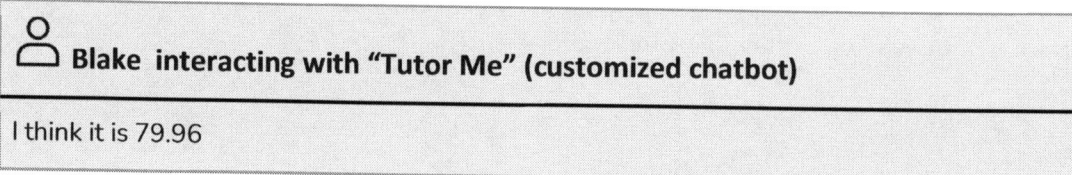

Blake interacting with "Tutor Me" (customized chatbot)

I think it is 79.96

"Tutor Me" (customized chatbot)

Let's calculate the tax amount correctly. The room rate is $99.95 per night, and the tax rate is 8%. How do you calculate 8% of $99.95?

The customized chatbot indicates that this is not the right answer, and provides more detailed guidance for Blake to solve this step of the problem. The problem is now down to a single calculation, which any student would likely answer correctly at this stage. Let's imagine, for illustration purposes that Blake does not know the answer and wants the chatbot to give them the solution directly.

Blake in "Tutor Me", a customized chatbot

I don't know, please give me the answer.

"Tutor Me" (customized chatbot)

I can help you figure it out. Let's start with the tax calculation again.

To find 8% of $99.95, you multiply $99.95 by 0.08. What do you get?

The chatbot still resists giving Blake the answer and instead remains supportive and breaks down the problem one more level to make it easier for Blake to solve.

And voila! This is what a Socratic AI tutor looks like in practice. In the back-end, "Tutor Me" is customized with instructions that indicate that it should not give students the solutions to the problems, it should remain supportive and understanding of their learning. This learning experience is, as illustrated, very different from that of the general ChatGPT conversation.

4. Using Custom GPTs Created by Others

Besides creating your own GPTs, you can also access those created by others. According to OpenAI more than 3 million GPTs had been created by January 2024 (just two months after announcing GPTs).[19] Not all of these GPTs are publicly available in the GPT store, but many are. After clicking on "Explore GPTs", you can access GPTs that have been built by other individuals or organizations. Furthermore, a myriad of GPTs on the web are tailored to educators. If you click on the tab "Education", you can see some of the most popular ones. We encourage you to spend some time exploring the many possibilities that GPTs built by others offer.

Figure 10.5 - Examples of Existing Custom GPTs for Education

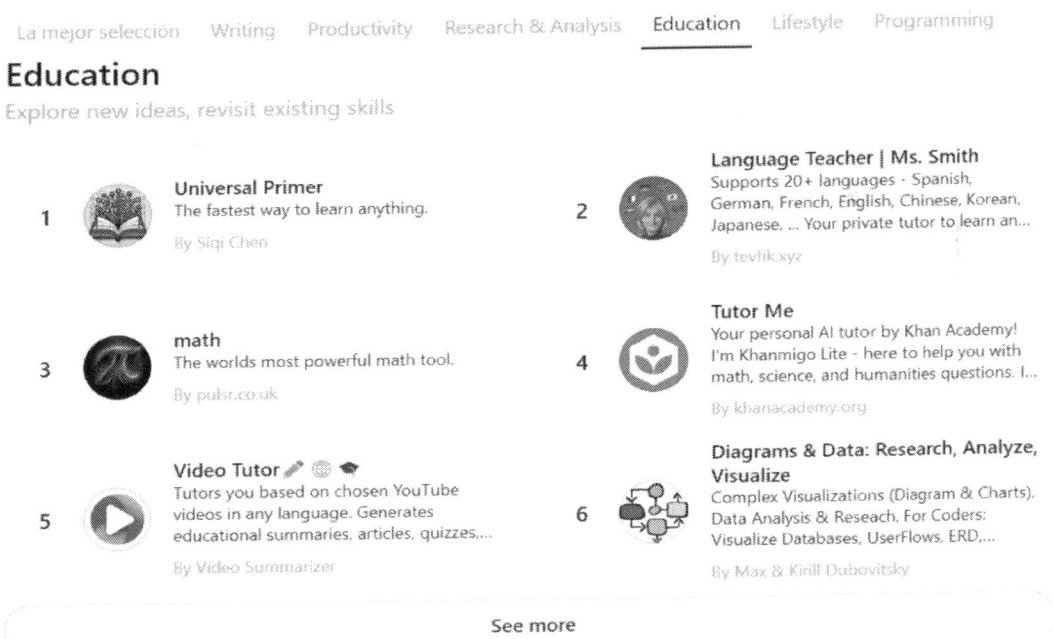

These GPTs come with built-in instructions to carry out specific tasks so you and your students don't have to build your own. For functional purposes like summarizing a paper, or building a presentation, we suggest you leverage GPTs built by others. For purposes that are specific to your course, like receiving feedback about a specific assignment or brainstorming practice questions, we suggest you build a GPT yourself.

To give you a sense of what custom GPTs are available in the GPT Store, we are listing a few that may be interesting to you below. The links to these bots are included in our companion site.

- **Tutor Me:** this GPT (mentioned in the previous example) was built by Khan Academy and is a lighter version of its paid AI tutor version, "Khanmigo." It serves the role of a personal AI tutor to help students with math, science, and humanities questions. It does not provide solutions but helps students learn how to solve them on their own.

Figure 10.6 - Khan Academy's Custom GPT "Tutor Me"

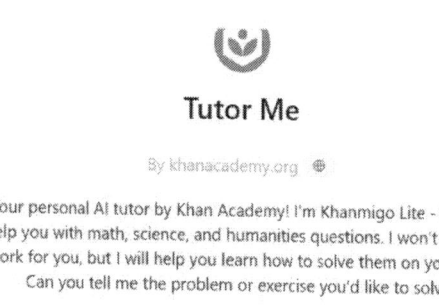

Tutor Me ∨

Tutor Me

By khanacademy.org ⊕

Your personal AI tutor by Khan Academy! I'm Khanmigo Lite - here to help you with math, science, and humanities questions. I won't do your work for you, but I will help you learn how to solve them on your own. Can you tell me the problem or exercise you'd like to solve?

Give me 10 practice problems!

How are you different than regular Khanmigo?

Message Tutor Me

- **Consensus:** this GPT serves as an AI-powered research assistant for scholars with access to over 200 million academic papers. It allows users to, among others, obtain answers based on research papers, and write draft content with citations.

Figure 10.7 - "Consensus" GPT

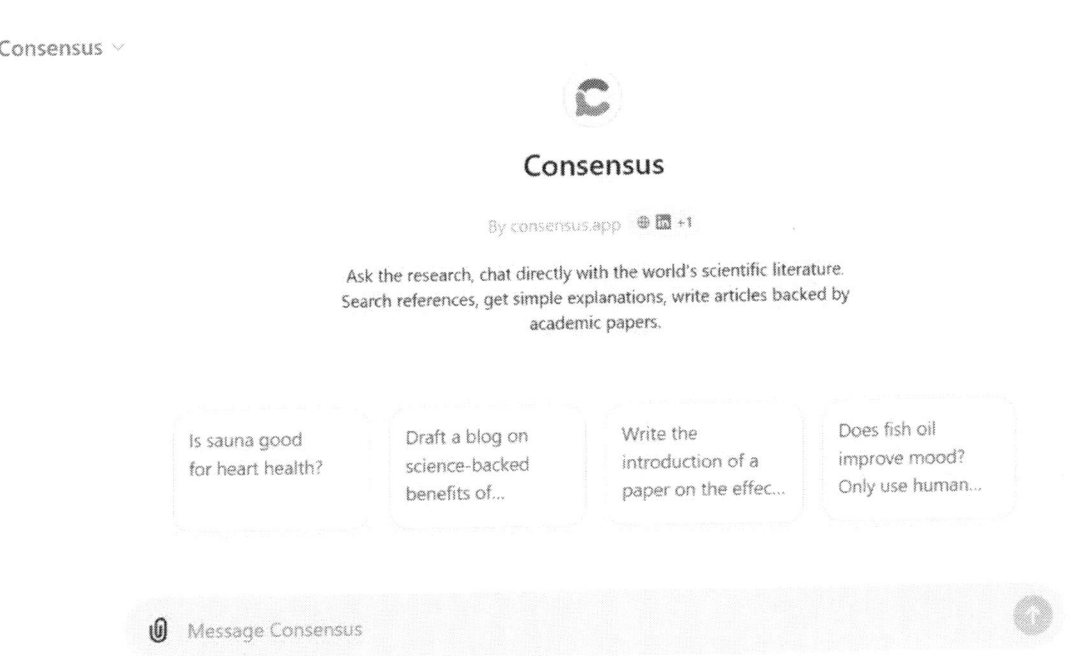

- **Wolfram:** this GPT was built by the makers of Mathematica. It allows users to access computation, math, curated knowledge & real-time data from Wolfram|Alpha and Wolfram Language.

Figure 10.8 - "Wolfram" GPT

Wolfram ∨

Wolfram

By wolfram.com ⊕ X

Access computation, math, curated knowledge & real-time data from
Wolfram|Alpha and Wolfram Language; from the makers of
Mathematica.

| Plot the intersection of x^2+y^2<1 and y>x with Wolfram... | get zodiac constellations visible from Chica... | Show me recent earthquakes in Indonesia | How has the dollar-euro conversion changed recently? |

 Message Wolfram

5. Leveraging OpenAI's API to Build More Advanced Chatbots

While custom GPTs are great solutions to most problems faced by educators, they may be too basic for more advanced users. For example, you may be interested in building a chatbot that is displayed in an interface other than ChatGPT, e.g. Slack, Google Colab, your university's learning management system, etc. If you find yourself too limited by the functionalities offered by Custom GPTs, you may be interested in exploring how to leverage OpenAI's API.

The process of building customized chatbots using OpenAI's API is beyond the scope of this book. If this is a topic you would like to understand in more depth, we are including a set of resources in our companion site. These resources include information about PingPong, a platform to create teaching-related bots powered by OpenAI's API created by Sharad Goel, Teddy Svoronos, and the team at the Computational Policy Lab at the Harvard Kennedy

School. Through this platform, faculty members can create their own bots in an environment where they can see the interactions that their students have with the platform, but OpenAI does not use the data for training.

We now move to examples from real educators using customized chatbots in their teaching. Some of the examples that are included below are built using OpenAI's API instead of building custom GPTs, but the general lessons would still be relevant if you wanted to build them using custom GPTs.

6. In Practice: Examples from Real Educators

The following section illustrates different ways in which real educators have made use of customized chatbots in their classrooms. We will continue building on the framework of the six possible student uses of AI that we presented in Chapter 8. For most of these uses, we are going to walk you through one or two examples of customized chatbots.

Table 10.2 - Customized Chatbot Examples

Student Use	Customized Chatbot	Designer
1- Get feedback	A writing editor for effective communication	Todd Rogers (Harvard Kennedy School)
2- Learn a new topic, concept or skill	A coding assistant for a computer science course	CS50 Teaching Team (Harvard University)
	A personal AI tutor for an economics course	Raj Chetty and Gregory Bruich (Harvard University)
3- Deepen their reflections	A reflective guide assistant	Cara Oneal-Radigan

	for middle schoolers	(formerly at a New Jersey school)
4- Find arguments and counterarguments	A debate simulation	Sharad Goel (Harvard Kennedy School)
5- Practice deliberately	A practice problems assistant for a statistics class	Teddy Svoronos and Sharad Goel (Harvard Kennedy School)
	A Negotiator Simulation	Ethan Mollick and Lilach Mollick (University of Pennsylvania)
6- Accomplish tasks	A process coach for PhD students	Vincent Cho (Boston College)

Given that the system prompts that configure these chatbots are generally very long, we have excluded them from this chapter. However, most of them are included in our companion site. We encourage you now to take a minute and go to our companion site, so you can read through the prompts as we walk you through the different chatbot examples.

Customized Chatbot #1: A Writing Editor

An example of a customized chatbot that is tailored to provide feedback to students is the chatbot created by Todd Rogers from the Harvard Kennedy School. Todd wrote a book with Jessica Lasky-Fink titled "Writing for Busy Readers" in which they provide science-based advice on how to write better to communicate more effectively to readers who skim. In his class, he often finds it challenging to provide enough personalized feedback in between sessions for students to practice the principles they have learned in class. Even in class, he admits it is hard to coach all his students.

To overcome this difficulty, Todd and his team built a customized chatbot that gives feedback to his students on the effectiveness of their writing. To tailor the chatbot's performance, they provided a checklist of the "Writing for Busy Readers" principles. They also provided example emails before and after they had gone through an editing process based on the guidelines from their book. The chatbot is used by his students to transform any text they submit into a more effective communication piece.

The tool they built was pedagogically powerful. Todd sometimes used the chatbot live during class, using examples provided by his students on the spot. This allowed him to illustrate the principles that he was teaching in a more interactive and engaging way than if he was showing an unrelatable hypothetical example. The chatbot is freely available and linked in the companion site.

To give you a sense of how his chatbot works, we submitted the three paragraphs above to it, and below is how it suggested we rewrite them.

Figure 10.9 - Example of Output from the Customized Chatbot Built by Todd Rogers

An example of a customized chatbot is the one created by Todd Rogers from the Harvard Kennedy School. He co-authored "Writing for Busy Readers" with Jessica Lasky-Fink, offering science-based writing advice.

Todd faced challenges in providing personalized feedback to students. To address this, he and his team built a chatbot to give writing feedback.

How the Chatbot Works:

- Uses example emails pre- and post-editing.
- Transforms student text into more effective communication.

Customized Chatbot #2.a: A Coding Assistant

CS50, Harvard's legendary Introduction to Computer Science course taught by David J. Malan, is one of the largest classes at Harvard, with over 600 on-campus students and thousands online. To support such a large cohort, David and his team created a customized AI tool to replicate the experience of having a tutor available 24/7. They acknowledged, "With this many students and teachers, providing a support structure has been a challenge."[20] The CS50 teaching team believes general AI tools like ChatGPT can be helpful but risk undermining learning by giving direct answers. To address this, they invested significant effort in developing their own AI tools on top of OpenAI's APIs.

The CS50 Duck, an AI-based chatbot, was a notable outcome of these efforts, inspired by the rubber duck debugging technique. David explains: "If you don't have a colleague, teacher, friend, or family member who knows more about programming than you, keep a rubber duck near your workspace. When you have a question or a bug, talk to the duck, walking through your problem verbally. While the duck won't quack back, the act of verbalizing your confusion often leads to a realization. As you explain the problem, you might suddenly understand what went wrong."[21]

The CS50 Duck aims to simulate a 1:1 teacher-student ratio by guiding students through their code, explaining errors, and suggesting improvements without giving direct answers. It can clarify lines of code, recommend style enhancements, and identify errors, fostering deeper understanding. Initial results among Harvard undergraduate students suggest that they have found these tools supportive and beneficial, enhancing their learning experience. About 75% of them used the AI tools more than twice a week, and about the same percentage found the tools either helpful or very helpful.

Customized Chatbot #2.b: A Personal AI Tutor

Raj Chetty and Gregory Bruich teach Economics 50, a popular course at Harvard titled "Using Big Data to Solve Economic and Social Problems." They collaborated with Evangelos Kassos and the Office of Undergraduate Education to create a chatbot, ec50.ai, like the one developed for CS50. Their goal was to help students understand course materials more deeply, ask questions about class concepts, and get assignment guidance. With around 500 students enrolled in Economics 50, providing personalized support is challenging, even with teaching assistants. The chatbot addresses various queries, including conceptual questions (e.g., what's upward mobility?), methods questions (e.g., how do I find the optimal depth in a decision tree?), and coding questions (e.g., how do I create an indicator variable in R?). The chatbot was built into the course's existing Slack workspace, which was the platform that students were already using to ask the teaching staff questions.

Here are some noteworthy aspects of how they configured the chatbot. First, they aimed to help students with exercises without providing direct answers. A common solution is to use the "Socratic method," where the AI asks questions to deepen understanding without revealing the answer. However, this can become frustrating and unproductive for students. To counter this, they programmed the chatbot to say, "I can't give you the answer," while also providing a helpful hint to encourage continued thinking and problem-solving.

Second, they wanted the chatbot answers to reflect the course materials and general approach. To achieve this, they specified that the chatbot should first look at the course materials for answers, and only use external information if no answer is found in the provided content.

Third, they wanted the chatbot to help students generate an unlimited number of new review questions on any topic covered in the course, along with detailed explanations of the correct answers. Critically, they wanted to encourage students to test themselves by keeping the solutions and explanations hidden until students had a chance to try the questions on their own. To achieve this, the chatbot was designed to function differently for review question generation than other uses of the chatbot. Students could prompt the chatbot with commands starting with a special prefix followed by the desired topic (e.g., "!externalities" would generate a practice question about externalities). In response, the chatbot would display only the questions (without solutions) in a main conversation, while keeping the suggested solutions and explanations hidden inside a threaded reply that could only be accessed with a click of the mouse.

Customized Chatbot #3: A Reflective Guide

Cara Oneal-Radigan, former Math and Science Teacher at a New Jersey school, also used customized chatbots as an opportunity to automate a time-consuming teaching task. In her case, Cara used to do a reflection exercise with her students at the end of every class. The exercise consisted of a "3-2-1 reflection activity" in which her students were asked to (i) list 3 things they had learned (ii) list 2 things they wanted to learn more about, and (iii) list 1 question they still have about the lesson. She then used the answers from her students as a form of "exit question" to understand how lessons were received and what skills or knowledge were mastered or needed more practice. Not only was this process manual, but she would sometimes get unhelpful answers from some of her students, and she regretted not being able to guide the reflections of enough of her students in an individualized manner.

With this in mind, she created a customized chatbot named "Class Pass Lass." Drawing on her previous manual experience, she tested and improved the chatbot by simulating multiple tests using answers from her past students. This allowed her to ensure that the assistant was helpful and stayed on topic without revealing answers, offering personalized support that teachers might find challenging to provide during the final moments of class.

Cara's full prompt is included in our companion site. If you read it, you will notice that Cara uploaded the lesson plan that she wanted her students to reflect on, so that the chatbot could check whether student responses were aligned or not with the lesson plan, and respond accordingly. She also included tips and example questions that the chatbot should give to the student if someone is struggling with the reflection exercise (e.g. "What is something your partner(s) said in your small group discussion that was interesting?"). One drawback of using a Custom GPT for this exercise is that Cara did not have access to student responses, and therefore could not leverage the reflections to understand better her students or to tailor her next lesson plans. This drawback could be addressed by building a customized chatbot leveraging OpenAI's API (see previous section).

Customized Chatbot #4: A Debate Simulation

Sharad Goel, professor of public policy at the Harvard Kennedy School, created a customized bot for his course, "The Science and Implications of Generative AI" (link to a free online version of the course is available on the companion site). Before a session on copyright and intellectual property, Sharad assigned readings about the New York Times lawsuit against OpenAI, which accused OpenAI of illegally training their large language models on millions of New York Times articles without permission.

Sharad then asked students to debate with the customized bot he created. He instructed, "You'll first be asked whether you generally support the New York Times or OpenAI. The bot will then take the opposite position and begin debating you." After the debate, students were asked if they had updated their position and to briefly explain why.

Some students found the customized bot to be a strong debate partner, appreciating its well-structured arguments, quick responses, and in-depth knowledge. Others felt it was weak and repetitive, noting that it sometimes agreed too readily, failed to introduce new arguments, or reiterated existing ones without offering creative counterpoints.

Overall, this strikes us as an excellent use of a customized chatbot to help students reflect and learn. One student reflected at the end "I think I still tend to be on the side of OpenAI but I am much less sure of my position and accept that there is some infringement of content creator rights that needs to be acknowledged and accounted for." Not a bad outcome for an experience that occurred before the student even set foot in the classroom!

Customized Chatbot #5.a: A Practice Problems Assistant

Teddy Svoronos, senior lecturer at the Harvard Kennedy School, has experimented with building multiple customized chatbots for his students. He starts by imagining what he would do for his students if he had unlimited time to dedicate to them.

With this in mind, he partnered with his faculty colleague Sharad Goel and his team at the Computational Policy Lab to create a custom bot called "StatGPT" for his introductory statistics course, which provided practice problems for students to master key skills. He did not have the bandwidth to create enough practice problems for his students but found this to be an essential way to learn the course materials. The system prompt he used to customize the chatbot is included in our companion site. The main things you should know is that the instructions include:

- The role that the chatbot is playing.
- The list of key skills learned in the course.
- Details on how the chatbot should interact with students.
- Instructions on the scenarios and context on which they should base their problems (e.g., policymaking in this case).

If you are checking the prompt on the companion site, you might notice that Teddy encourages the bot repeatedly to double-check its work and to produce step-by-step calculations. This generally leads to better results from ChatGPT. Teddy built the chatbot to be integrated with Slack, given that this is the technology he uses to communicate with his students, so he specified this within his prompt. Overall, the experience was positive for students, with 68% of them reporting StatGPT to be effective or very effective for their learning.

In his experimentation, Teddy has found that dedicated chatbots with specific, narrow purposes tend to be more effective than general-purpose bots that students can interact with however they like. This is why he prefers to create various bots for different purposes even if they are going to be used in the same course. For example, he co-taught a class called "The Science and Implications of Generative AI" and built two bots for the students. The goal of the first one was to ask questions and evaluate the correctness of student answers about the course syllabus. The goal of the second one was to ask students to explain concepts from the course in an accessible manner and then provide them with feedback that draws on class materials, suggesting ways to better explain the ideas. Both prompts are available on our companion site.

Customized Chatbot #5.b: A Negotiation Simulator

Lilach Mollick and Ethan Mollick used a Custom GPT to do a negotiation role-play exercise with their MBA students. They had two goals in mind when they crafted the prompt. First, to help the chatbot understand its task. To achieve this, they included step-by-step instructions and examples in their prompt. For example, they specified lessons and tactics that a well-versed negotiator would consider in a negotiation. Their second goal was to ensure that the chatbot created a positive and supportive experience. To achieve this, they instructed the chatbot to present only one question and one choice at a time. They also ensured that if a student lost focus or seemed stuck, the chatbot would provide hints. Additionally, they instructed the chatbot to provide feedback organized around topics.

They recommend debriefing the experience after assigning the exercise to their students. They highlight that it is important to explore both how this exercise impacted their learning, but also how successful the chatbot was in creating the simulation.

Customized Chatbot #6: A Process Coach

Vincent Cho, associate professor of education at the Lynch School of Education and Human Development at Boston College, used to help his PhD students refine their problem statements for their dissertations. The process of doing so however was very time-consuming. He would ask students to fill in a worksheet identifying key elements of a problem statement (e.g. a gap in the research, a purpose statement, a theoretical framework, etc.) and then provide individualized feedback.

To automate this process, he built a customized chatbot that guided students through the same process but did it in a digestible way by breaking up the information into bites and sharing it with his students at the appropriate stage in the process, while providing a space to ask questions and get quick feedback and reactions on their different ideas on the spot.

The full system prompt is included in the companion site. The components of the prompt include:

- Role
- Context
- Task
- Rules to follow
- Process steps
- Wrap-up and next steps

If you are checking the prompt, notice how detailed some of the instructions for each step are, e.g. "Explain how choosing different verbs, such as explore, evaluate, describe, or explain could shape the direction of their study." Most likely, this level of guidance was possible thanks to the level of experience Vincent Cho had in doing this exercise manually with his students in the past. If you don't have experience doing an exercise with students manually before building a customized chatbot, we encourage you to simulate student answers or to ask some former students of yours to play with the chatbot. It is hard to know what feedback or guidance you need to provide before you have tested a process!

After trying this with his Ph.D. students, they showed appreciation for the reassurance they received from the chatbot when they were unsure about where to start or how to develop their ideas further.

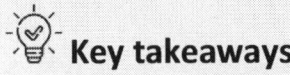 **Key takeaways**

- **Ease of Customization:** You don't need to be a tech expert to build customized chatbots with ChatGPT. This chapter showed you how straightforward it is to tailor interactions and improve efficiency in your teaching, even if you have no coding experience.

- **Personalized Instruction:** Customized chatbots offer you the chance to create a highly personalized and relevant learning experience for your students. By aligning chatbot interactions with your specific course materials and instructional needs, you can better support your students.

- **Iterative Improvement:** Building Custom GPTs (the ChatGPT version of a customized chatbot) is an iterative process. You will continually test, receive feedback, and adjust your bots, ensuring they effectively meet your instructional goals.

- **Practical Applications:** Throughout this chapter, you have seen real-world examples of how educators like you use customized chatbots. From creating quizzes and providing writing feedback to tutoring and guiding reflections, these examples illustrate the diverse ways you can enhance your teaching and your students' learning experiences. So go ahead and experiment with creating one for you and/or your students!

Part IV – In Closing

Chapter #11 - Beyond ChatGPT

This book aims to be an introductory and practical guide for educators who want to understand how to use ChatGPT to teach more effectively and to improve their students' learning. We made a conscious choice to narrow the scope of the book to ChatGPT but would now like to give you a sense of the myriad of AI tools available online that could also help you in your teaching. Since the field is rapidly changing, some of these tools might not be available by the time you read this chapter, others might have changed or added features, and some new important tools might have emerged. Nevertheless, we hope this chapter gives you a sense of some of the possibilities beyond the chatbot text-based interface of ChatGPT.

1. Other Chatbots

Chatbots are powered by Large Language Models (LLMs), a technology that can respond to and generate human-like text based on natural language processing. In the case of ChatGPT, it is powered by an LLM called GPT-4o (the latest model) created by OpenAI. As you have seen with ChatGPT, these AI tools can engage in dialogue, answer questions, write content, and perform other similar language-related tasks.

Other conversational AI tools that you could explore or use in ways similar to ChatGPT include the following:

- **Claude:** Anthropic's Chatbot which offers similar functionalities to ChatGPT but claims to place additional emphasis and focus on user safety. It also offers a paid and a free version.
- **Gemini:** formerly known as Bard, Gemini is Google's Chatbot and also offers a free and paid version.
- **Llama:** Made by Meta (formerly Facebook), it is among the most popular open models in the market available for free.

In our experience, right now the functionalities of these LLMs are similar enough that it is not worth spending much time comparing them. If you are curious to go beyond ChatGPT, just pick one and experiment with it.

2. Specialized AI Tools Beyond General Purpose Models

Beyond general-purpose models like the ones we have explored so far, there are many AI tools that optimize their models to serve specific functions. The field is very large and growing rapidly, but we include a table below with suggestions of some that you can start experimenting with now. We then briefly describe some of the tools. A full treatment of them is beyond the scope of this book.

Table 11.1 - Specialized AI Tools

Function type	Role	Example of AI Tool
Search	Integrates AI into web searches	Perplexity AI
Research	Finds literature review resources with links to sources	Consensus, Elicit, Scholar AI, SciSpace
Image Generation	Creates images based on a set of instructions	MidJourney, Stable Diffusion
Presentation Aides	Creates, edits, and summarizes slide decks	Microsoft Copilot, Google Gemini
Speech Coaching	Provides feedback on oral skills	Yoodli Ai
Coding	Supports with coding production and debugging	GitHub Copilot, Google Colab AI, TabNine

Productivity	Enhances user productivity by automating tasks and processes	Gamma AI (slide building), Grammarly (grammar checks), Otter AI (meeting transcription), Canva (design), Clockwise (scheduling), Zapier (general automation)

Search

In its early days, ChatGPT could not answer questions related to events that occurred after its training cutoff date. For example, it could not answer questions such as "What was the result of the Portugal vs. Turkey soccer game yesterday at Eurocup 2024?" Now that ChatGPT added web searching to its capabilities, it can search the web for queries like this one, but this is not its strength. Several AI tools have been developed in this space, and traditional search engines like Google are incorporating AI in their searches.

As of the time of writing, one of the most popular apps in this category is "Perplexity AI" which brands itself as a "free AI-powered answer engine that provides accurate, trusted, and real-time answers to any question."[22] In contrast to some of the other AI platforms (including ChatGPT), Perplexity provides citations in all its responses, so you can click the link to the source if you want to dig deeper or verify the accuracy of the information.

Research

If you are interested in using AI to help you search the academic literature on a certain topic, AI tools specialized in this task are likely to be much more helpful and accurate than ChatGPT. The sorts of questions you can ask include: Does exercise improve cognition? Do cash transfers reduce poverty? Are statins effective for the elderly? There are several apps in this field, including

Consensus, Elicit, Scholar AI and SciSpace. Please note that several of these apps can be accessed both in a web browser and as a Custom GPT in the GPT Store.

For the sake of brevity, we will describe Consensus, which brands itself as an "AI search engine for research papers. You can think of it like Google Scholar meets ChatGPT".[23] You can ask a question such as "Can CBT improve anxiety?" and Consensus will display a list of the main papers in that literature, and for each of these papers a summary of the key findings, the type of research design that was used, and a study snapshot that describes study features such as sample size and population studied. It will also synthesize the findings across the papers. Moreover, you can ask questions about the papers cited using their "Copilot" feature.

Image Generation

The paid account of ChatGPT allows you to create images from text. This can be helpful in your teaching, research, and day-to-day work. Please note that as of the time of writing, these image generators were not good at incorporating text in their images; the text was frequently displayed with typos. If you want to go beyond ChatGPT's image generation capabilities, popular alternatives include Midjourney and Stable Diffusion.

Presentation Aides

As described in chapters 3 and 4, the field of apps that can create or edit slides with the help of AI is evolving quickly. Companies that offer software to create slides are actively incorporating AI into their apps. For example, Microsoft Copilot allows you to create and edit PowerPoint slides with the assistance of ChatGPT, and Google Slides allows you to do the same with the assistance of Google Gemini. Moreover, these platforms allow you to summarize and ask questions about an existing slide deck. Both of these services require a paid subscription, and their uses extend to other apps in their office suite (e.g., Word, Excel, Google Docs, etc.). Our current assessment is that these tools are impressive to see in action but not yet ready for primetime. We suspect that this will change and that very soon many of us will be creating presentations jointly with AI in a way that will become very natural (i.e., inside the app without the need to go to a chatbot). The same will be true for writing documents, analyzing data in spreadsheets, and many other functions. See the companion site for tutorials on some of these tools.

Speech Coaching

In Chapter 8, we saw how Hadar Sachs, a student at the Harvard Kennedy School, used ChatGPT to practice and get feedback on a presentation she was preparing for one of her classes. ChatGPT gave her feedback on the structure of her arguments, her language, and other aspects of her speech. She could have taken this process beyond ChatGPT by obtaining feedback on her delivery from a specialized tool that can analyze her speaking pace, intonation, and emphasis.

Such tools already exist. Allison Shapira, who teaches a course in the Arts of Communication at the Harvard Kennedy School and leads a company that provides training on public speaking and presentation skills (see companion site for details), has encouraged her students to use a tool called Yoodli.ai to practice their speeches for her course. The tool provides AI-driven feedback on content and delivery. The benefit of using a tool like this is that it provides her students with many more opportunities for feedback than she could provide on her own given time constraints. In her case, she partnered with the company to train a version of the model on her public speaking book, ensuring that the advice her students received was tailored to her methodology.

Coding

One of the industries most disrupted by AI is coding. Modern programming is now assisted by AI, and AI-powered tools such as GitHub Copilot, Google Colab and TabNine have revolutionized the way developers write code. These tools can autocomplete code snippets, suggest entire functions, and even detect and correct errors in real time.

Hong Qu, who teaches data visualization at the Harvard Kennedy School, encourages his students to use prompts to generate Python code in Google Colab. This approach helps them learn Python syntax and build confidence in their ability to write or prompt AI to write code without constantly referring to sites like Stack Overflow. He emphasizes, "I am not teaching a Computer Science (CS) programming class, but I don't encourage students to use AI for coding until after I have taught the fundamentals of Python. AI serves as a coding copilot for data processing automation, ultimately aiding in visualization. In a pure CS class, I would introduce AI much later."

Productivity

There are numerous AI-powered apps designed to boost productivity by automating tasks we typically do manually. While this topic could fill an entire book, we briefly explore a few use cases that might interest you. These apps can either directly aid your teaching and learning or free up your time, allowing you to focus on more valuable activities in both your professional and personal life.

- **Grammarly (grammar checks):** Grammarly provides real-time grammar, spelling, and style suggestions to enhance the clarity and accuracy of your written content.
- **Otter AI (meeting transcription):** Otter AI transcribes meetings and lectures automatically, making it easier to review and share key points.
- **Canva (design):** Canva offers tools for creating visuals for lectures, handouts, and social media, supporting the development of engaging content.
- **Clockwise (scheduling):** Clockwise manages your calendar by scheduling meetings and optimizing time blocks, reducing the need for manual scheduling.
- **Zapier (general automation):** Zapier connects apps and automates workflows, helping to minimize repetitive tasks and streamline processes.

Finally, many of the apps you use daily are integrating AI features. For example, Zoom and Microsoft Teams now have AI assistants that provide quick recaps if you join a meeting late, and summaries of key points and action items at the end. Similarly, Google Docs uses AI to suggest edits and improvements to your writing, and Gmail's AI-powered Smart Compose helps you draft emails faster by predicting text as you type.

3. Modalities Beyond Text

In most of the examples highlighted in this book, the input to the AI tool has been in the form of text (usually as a written prompt) and the output has also come in text format. However, it is becoming increasingly common to interact with AI tools in modalities other than text. In fact, the "o" in ChatGPT 4o stands for "omni." This name reflects the model's enhanced capabilities, allowing it to handle a wide variety of input types, including text, voice, images, and video. More broadly, other AI tools are also becoming multi-modal, and this seems like a general trend in the market. Table 11.2 describes some of the many possibilities.

Table 11.2 - Modalities Beyond Text - Examples

Input	Output	Example of AI Tool
Text	Text	ChatGPT, Claude, PingPong
Text	Image	DALL-E, Midjourney
Text	Voice	ElevenLabs
Text	Video	HeyGen, OpenAI's Sora
Image	Text	ChatGPT (Vision), Claude (Vision)
Voice	Voice	ChatGPT

Data	Charts, Tables, etc.	ChatGPT (Advanced Data Analysis)
...	...	...

A full treatment of all the combinations of interacting with AI tools aside from text is beyond the scope of this book, but below we describe two of the possibilities.

Image to Text

If you have never used the vision capabilities of ChatGPT, we ask you to pause what you are doing and find a picture of a cooked meal (from your camera roll or on the internet). Now attach that picture to ChatGPT and ask it to provide the recipe for that dish. It won't be able to provide the exact recipe but if your experience is anything like ours, you will be shocked at how good it is at this task!

Apart from this kind of fun use of ChatGPT's vision capabilities, you might be able to imagine some applications more relevant to education. For example, you can upload a screenshot of a table and ask ChatGPT to translate it into an editable table that you can copy and paste into a document. We also heard from a teacher who used ChatGPT to upload images of "sticky notes" from a working session she had with parents, and she then asked ChatGPT to translate the images into text and categorize the content of the notes!

Other examples abound. In subjects like biology, geography, or art history, students could upload images or diagrams (e.g., a cell structure, a map, or a painting) and ask ChatGPT to analyze or explain specific elements. Art and design students could upload their projects or works in progress and receive constructive feedback on composition, technique, and other artistic elements from ChatGPT. Finally, students can upload images of complex math problems, equations, or geometric figures and ask ChatGPT to help them with a step they are stuck on. And yes, unfortunately they could also ask ChatGPT to solve the problem for them, which will hinder their learning.

Voice to Voice

An increasingly popular way of interacting through ChatGPT is through voice. The interaction is similar to the one you would have with traditional assistants such as Siri or Alexa. You start speaking and when you pause for a few seconds, ChatGPT will respond by voice. This is a particularly attractive use if your hands are occupied (e.g. when you are driving or cooking). And because you are interacting with ChatGPT, it remembers what you said before, so it almost feels like a normal conversation (though with a slightly longer lag). Furthermore, the conversation gets registered as text and becomes available in your chat history, just like any other conversation you have with ChatGPT. If you have never tried it before, we encourage you to pick up your phone, download the app and try it now.

Apart from you or your students interacting with ChatGPT via voice when it's more convenient to accomplish tasks traditionally done through text, you might be able to envision other ways in which this modality could be useful in education. For example, imagine a classroom where you as the instructor could occasionally pause and speak to ChatGPT to ask a question, inquire about an example, or assume a persona to debate with (e.g., a historical figure). Another possible use of the voice feature that may be attractive to language instructors is to have students engage in spoken conversations with ChatGPT in the target language, practicing pronunciation, fluency, and conversational skills. The voice capabilities of ChatGPT could also be helpful for students with visual impairments or learning disabilities who can use voice commands to interact with ChatGPT, asking for explanations, summaries, or assistance with course material. These are only some of the many possibilities.

4. Education-specific AI Tools and Communities

This section will cover AI tools that have been built specifically for educators and learners. This space is growing very rapidly. Here we will briefly describe a few in the spirit of helping you discover some we might want to explore. We have not tested all of these tools, so please explore them at your own risk.

- **MagicSchool:** a platform that consolidates different AI-powered apps to help educators automate administrative tasks and personalize learning experiences for students.
- **Quizlet:** A study tool that helps students learn with flashcards, quizzes, and personalized study plans. It has now incorporated AI and made it easier to produce flashcards and other items.
- **Khanmigo:** an AI-powered tutoring tool developed by Khan Academy to provide personalized learning assistance and support to students.

We cannot end this section without acknowledging that there are also AI-driven apps and websites that can be used by students to learn but also to short-circuit the learning process. For example, AnswersAI allows users to screenshot a problem, and the tool gives them the solution. Similarly, apps like Socratic and Photomath enable students to take pictures of math problems and receive step-by-step solutions instantly. While some of these capabilities are also available on ChatGPT, some of these specialized apps might make it a bit easier.

5. Where Things Are Heading

It is challenging to speculate where the AI world is heading and how future changes will affect education. But to plant some seeds in your mind, we highlight two trends we see and thoughts on how they might affect your teaching:

- **Emergence of Agents:** So far, most uses of AI have generally been limited to one task at a time. However, the use of agents will become increasingly common, where you give an AI an objective, and it performs multiple tasks to achieve that objective. Imagine you could issue a command like, "Please help me support the students who are at risk of failing my course." The AI tool could sift through student data, determine which students are at risk based on exam scores, attendance records, etc., and draft personalized emails for you to send to these students.

- **More Personalized AI:** Currently, AI knows very little about you, but soon you could enable it to know much more (for example, that you are writing a book or going on a trip tomorrow) and adjust to your circumstances much better. This is certainly where Apple's AI strategy is heading. Imagine as a faculty member, your AI assistant could remind you about grading deadlines, suggest resources for your next class session based on current events, or even schedule your meetings around your peak productivity times.

We hope this gives you a sense of how things might evolve. Changes will continue to happen rapidly, and as educators, part of our job will be to adapt to these changes and leverage them to improve our teaching and help our students learn better.

Key takeaways

- **Other LLMs:** ChatGPT is not the only game in town. You can use other Chatbots powered by other Large Language Models (LLMs). These include Claude, Google

Gemini and Llama.

- **Specialized Apps:** There are many specialized apps that might do a better job than the general-purpose ChatGPT technology. They specialize in tasks such as search, research, image generation, presentation aides, speech coaching, coding, and productivity. If you are interested in using AI for any of these tasks and are not happy with ChatGPT or your current solution, it might be worth exploring some of these apps.

- **Modalities beyond text:** AI tools are quickly evolving to handle various input and output modalities, including text, voice, images, and video. You will be increasingly likely to speak rather than type when communicating with AI. This might open up some uses both inside and outside your classroom.

- **Education apps:** There are several specialized education apps that leverage AI. It might be worth exploring some of them to help advance your learning goals. It is also good to be aware that some of them can be used by your students to learn or to bypass the process of learning when doing your assignments.

Chapter #12 - Conclusions

This last chapter gives you some pointers on how to bring together everything you have learned and suggests ways to plan your next steps. Before proceeding further, take a few minutes now to write down the main ideas you have gathered from the book. Which lessons resonated more strongly with you? What new ideas do you want to implement in your courses? How can ChatGPT be used in your teaching to enhance your students' learning? If you kept notes while reading, this is a good time to update them with any new thoughts. Doing this will deepen your understanding and make it more likely you will use these ideas in your teaching.

Key Ideas

We take a step back to remind you of some of the key ideas in the book that we hope can guide you in leveraging ChatGPT in your teaching.

1. **Focus on Learning:** Your goal is to help your students learn. ChatGPT and other AI tools are a means to achieve this.
2. **Guiding Principles:** Use AI to augment your capabilities, remember that you are the expert, treat AI as a conversation partner, and experiment continuously.
3. **Assistance for educators:** ChatGPT can assist in various aspects of teaching, including improving existing classes, preparing new sessions, designing pre-class work, enhancing in-class activities, and creating assignments. Think of ChatGPT as a 24/7 assistant.
4. **Assistance for students**: ChatGPT can significantly aid student learning but can also hinder it if misused. When used correctly, it can serve as a personalized tutor with infinite patience. Students can use it for feedback, learning new topics, deepening reflections, finding arguments and counterarguments, practicing deliberately, and completing tasks.

5. **Evolving Skills and Competencies**: Generative AI is changing the skills and competencies our students will need. This should force us to re-examine what we teach, how we teach, and how we assess our students.

In the appendix below we compiled the key ideas from each part of the book. Our advice is that you highlight 3-5 ideas to start your experimentation with AI. You can always come back for more.

Next Steps

We wrote this book with the hopes of nudging, inspiring, and empowering you to experiment with ChatGPT to enhance your teaching and the learning of your students. Hopefully, you arrived at the end of this book with some ideas that you are excited to implement. We will leave the specifics to you, but want to nudge you to think intentionally about how to take what you have learned in this book forward. To do so, below are some ideas that you can consider as you plan your next steps.

- **Start experimenting:**
  - Choose a use of ChatGPT that you believe will add value to your teaching and that you think is feasible to implement.
  - Start a chat with ChatGPT. Have a conversation.
  - Iterate with various prompts and guide ChatGPT towards what you want to achieve. If it does not seem to be working, break down your project into steps and try getting assistance one step at a time.
  - Reflect on what is working and what isn't.

- **Learn with and from your colleagues:**
  - Share your ideas, learnings, and reflections with your colleagues.
  - Ask your colleagues how they use ChatGPT (and other AI tools) and what they are learning from it.

- **Make a plan for how to integrate ChatGPT into your classroom:**
  - Think about one way you may want your students to use ChatGPT in your course.
  - Consider how you may discuss the use of ChatGPT with your students.

- **Finally, be patient.** It might take some time experimenting before you see any large benefits. But once you do, you will start seeing use cases everywhere!

In Closing

This book has been nourished by the contribution of dozens of educators and students who have shared with us how they are using ChatGPT. Learning from others is key to any learning process. If you come up with some ideas that you think would be of interest to other educators, we would be grateful to hear from you. You can share your ideas with us on the companion site and we will help you share them with others (giving you full credit, of course). We will also welcome your feedback about this book, as you consider or implement some of these ideas, so we can all continue learning about this rapidly advancing technology.

Finally, let's remember that technology is merely a vehicle to help us achieve our goals as educators: nurturing our students' learning, growth, and development. ChatGPT is one of many tools at our disposal to facilitate this process. In these uncertain times, it can be challenging to distinguish what is genuinely useful. Yet, one thing remains certain: devoted and passionate educators are the heart and soul of a student's learning journey. We wish you all the best on this journey and hope that what you have learned in this book will prove to be helpful along the way.

Appendix – Key Ideas

This appendix contains the key ideas from each part of the book, so you can have them in one place for easy reference.

Part I - Summary of Key Ideas

Generative AI is here to stay:

- Generative AI technologies are not likely to stop evolving any time soon. Start using the technology now and adapt to the coming changes rather than try to catch up years later.
- Whether we like it or not, generative AI will be a part of our students' experiences both in and beyond the classroom. This affects how you teach and how you think about their learning.

How to approach using ChatGPT in your teaching:

- Your main goal is to help your students learn. While it's important to understand some aspects of ChatGPT, you don't need to become an expert in prompt engineering to use it effectively.
- This book aims to provide you with practical ideas and strategies to integrate ChatGPT into your teaching, to empower you, and to ultimately inspire you to teach more effectively using ChatGPT.
- OpenAI uses the aggregate information from our conversations with ChatGPT to improve the performance of their models. Unless your institution has arranged your access to a version with privacy protection, we suggest not exposing sensitive information, such as student names or grades, to the system.

Three pedagogic principles to guide your use of ChatGPT:

- Be student-centered. Recognize that teaching is not the same as learning.
- Plan for active learning. Students learn best when they actively engage with the material through processing, application, inquiry, and interaction.
- Begin with the end in mind. Start by asking what you want your students to achieve, rather than what you will "cover" in class.

Four AI principles to guide your use of ChatGPT:

- Use AI to augment your capabilities. The question is not whether ChatGPT is better than you, but rather whether you and ChatGPT together can be better than you alone.
- You are the expert. You understand the accuracy and relevance of ChatGPT's output and your students better than anyone else.
- Treat AI as Your Conversation Partner. Provide feedback, ask follow-up questions, try different ways of asking the same question, and refine your queries.
- Experiment. Experiment. Experiment.

Four prompting principles to guide your use of ChatGPT:

- Consider the formula "Task" + "Instructions" + "Context" to think about all the key elements that may be required in a conversation with ChatGPT.
- When in doubt, ask ChatGPT what information it needs to do the task correctly.
- When you believe ChatGPT made a mistake, tell it so.
- Break complex tasks into smaller, simpler components.

Part II - Summary of Ways You Can Use ChatGPT

Some key benefits of using ChatGPT in your teaching:

- Draw inspiration from sources or ideas you would have otherwise not considered.
- Save time by automating tedious tasks or simply making you more productive.

- Improve your teaching by generating engaging activities, personalizing your teaching to your students, and getting instant feedback from ChatGPT.

Six ways of improving an existing class with ChatGPT:

- Improve your slides: Upload your slides to ChatGPT and ask for feedback.
- Update your class activities: Upload your slides to ChatGPT and ask for ideas of new activities you can add, or upload an existing class activity and ask for an updated one.
- Generate explanations, examples, or analogies: Instruct ChatGPT to provide you explanations, examples, or analogies that will match the background (level of expertise, interests, etc.) of your students.
- Improve your class plan: Upload your class plan to ChatGPT and ask for feedback.
- Summarize student feedback: Ask ChatGPT to summarize key themes in your course evaluations and make suggestions for improvement.
- Get feedback on your teaching: Ask ChatGPT to analyze a transcription from a recording of one of your classes (assuming you have privacy protections to do so) and provide suggestions for improvement.

Four ways to prepare for a new class using ChatGPT:

- Design a class plan: Use ChatGPT to brainstorm or ask for feedback on your learning goals, learn about the topic at hand, identify resources you could read to prepare, or suggest a time plan or way to structure your class.
- Generate engaging in-class activities: Ask ChatGPT to design activities that generate active learning, such as role-plays, simulations, and group discussions.
- Prepare slides and other materials: Draft an outline first and then produce the slides based on this outline. ChatGPT can help create the slides and ensure they are clear and concise.
- Plan your in-class assessment: Use ChatGPT to create exit tickets, one-minute papers, and other formative assessments to gauge student understanding during class. This

allows for real-time adjustments to your teaching.

Five ways to design pre-class work using ChatGPT:

- Identify or produce useful resources: Use ChatGPT to identify books, articles, videos, and other resources related to your class (this is an area where ChatGPT is particularly prone to hallucinating). Alternatively, use it to produce new resources such as caselets.
- Generate questions for students: Use ChatGPT to brainstorm ideas for questions, contextualize the questions to a relevant setting, adjust existing ones, or review your questions.
- Collect student answers: Use ChatGPT to produce a document (such as Word or Google Form) to collect the answers of your students to the pre-class exercise.
- Analyze student answers: Use ChatGPT then to analyze both quantitative (e.g. generate graphs) and qualitative answers (e.g. find main themes).
- Plan your class accordingly: Use ChatGPT to identify topics that seem mastered and those that do not and adjust the class accordingly, or find quotes from students to reference in class.

Several ways to use ChatGPT during class:

- Use ChatGPT to collect and summarize student questions or key takeaways during class. This can enhance participation and provide immediate feedback to address confusion.
- Use ChatGPT as a live teaching assistant, as a group work assistant, to simulate debates, create interactive case studies, or provide enhanced accessibility. These uses are based on our speculation of what will be possible.

3 ways to assess or grade your assignments:

- Design or improve an assignment or exam. ChatGPT can give you feedback on an existing assignment, help you tweak or change the context of an assignment, or assist you in creating new assignments.

- Design grading rubrics. ChatGPT can help you brainstorm learning goals for your assignments, assist you in creating rubrics based on those goals, and provide you feedback on your rubrics.
- Provide feedback to your students and get help to grade their work. While using ChatGPT in this way can help you save time and perhaps provide more extensive feedback, consider the drawbacks involved before deciding whether to follow this path.

How to optimize your use of ChatGPT:

- Iteration is key.
- Useful contextual information to provide includes: the title and goals of the course you are teaching; the background of your students (especially in the topic at hand); any teaching/pedagogic principles to keep in mind; specific details of the task at hand (e.g. format of questions to be generated, time available for making changes...).
- Assess and judge whether the response from ChatGPT is accurate and useful to you. Not all will be valuable, but some may, and the cost of obtaining it is almost negligible.
- Remember that ChatGPT can generate Microsoft documents such as Word or PowerPoint. To create Google documents such as Docs, Slides, or Forms, you can ask for a Google Apps Script.
- Remember also that it can easily reformat existing content or data (e.g. create tables).

Part III - Summary of Ways Your Students Can Use ChatGPT

How to think about the impact of ChatGPT on your students:

- Different uses of ChatGPT can lead to different degrees of learning. There is no such thing as "ChatGPT is good for learning" or "ChatGPT is bad for learning".
- ChatGPT can be a complement to your teaching, enabling the opportunity of your students to get closer to a 1:1 personalized learning experience.

- The ubiquity of AI calls for redefining educational standards. AI elevates everyone's performance to a new minimum level.

Three concerns about students using ChatGPT and how to think about them:

- Students might cheat. Considerations: (i) No reliable way to detect AI-generated text; (ii) Increasingly difficult to document AI involvement; (iii) You can enforce lax, guided, or restrictive AI policy, but enforcing it can be challenging; (iv) You can rely more on assessment methods that are robust to AI (e.g. in-person tests).
- Students might receive false or biased information. Considerations: (i) Foster critical thinking; (ii) Think about the counterfactual; (iii) Assess areas where ChatGPT is most likely to mislead students; (iv) Assess costs of inaccurate information; (v) ChatGPT technology will get better.
- Students might become lazy. Considerations: (i) Assess which competencies in your course are still relevant and which are not; (ii) Assess how the use of ChatGPT affects the acquisition of those still relevant.

Six ways your students can use ChatGPT to learn:

- Get feedback. Students can obtain instant individualized feedback on their work (e.g. essays, presentations, etc.), which can have significant benefits for learning.
- Learn a new topic, concept, or skill. Students can treat ChatGPT in some ways as a private tutor with endless patience and 24/7 availability, which can empower them to take ownership of their learning.
- Deepen their reflections. Students can use ChatGPT to articulate their thoughts or deepen their reflections, activating metacognition.
- Find arguments and counterarguments. Students can use ChatGPT to brainstorm possible arguments and counterarguments or to simulate a debate, considering alternate viewpoints.
- Practice deliberately. Students can use ChatGPT to generate opportunities for

practicing and applying their knowledge such as producing test questions on a given topic.

- Accomplish tasks. Students can use ChatGPT as a productivity tool to free up time they can decide to spend in more valuable ways. The impact of this use on their learning is uncertain.

Nudging your students to use ChatGPT:

- Students generally welcome the opportunity to experiment with different uses of ChatGPT.
- Student learning can benefit significantly if teachers encourage and establish "guardrails" to their use of ChatGPT. It is helpful to think about which student uses of ChatGPT can benefit more from your guidance or nudging in a way that is most conducive to learning.
- When designing activities where you expect your students to use ChatGPT, it is useful to consider the degree of guidance you will provide and the degree of monitoring you will do.
- Building AI literacy can be a learning goal on its own to better prepare students for the complexities and demands of the modern world. Some educators encourage their students to experiment with different generative AI tools and learn key skills like prompt engineering.

Building customized chatbots

- Customized chatbots can help tailor ChatGPT's technology to your specific instructional needs. Compared to a general ChatGPT conversation, customized chatbots can help you automate repetitive tasks, share your "model" with others, and address privacy concerns.
- Two key approaches to building customized chatbots. The simplest one is building custom GPTs in the ChatGPT interface. The advanced one is building a chatbot using

the OpenAI API.

- As an educator, your customized chatbots could serve to support each of the six ways in which students can use ChatGPT to learn.
- It is helpful to narrow down the role of your chatbot, provide step-by-step instructions on how it should address the needs of your students, and test the chatbot repeatedly to iterate and refine its configuration before sharing it with students.

Part IV – Summary of In Closing

- Other LLMs: ChatGPT is not the only game in town. You can use other chatbots powered by other Large Language Models (LLMs). These include Claude, Google Gemini, and Llama.
- Specialized Apps: There are many specialized apps that might do a better job than the general-purpose ChatGPT technology. They specialize in tasks such as search, research, image generation, presentation aides, speech coaching, coding, and productivity. If you are interested in using AI for any of these tasks and are not happy with ChatGPT or your current solution, it might be worth exploring some of these apps.
- Modalities beyond text: AI tools are quickly evolving to handle various input and output modalities, including text, voice, images, and video. You will be increasingly likely to speak rather than type when communicating with AI. This might open up some uses both inside and outside your classroom.
- Education apps: There are several specialized education apps that leverage AI. It might be worth exploring some of them to help advance your learning goals. It is also good to be aware that some of them can be used by your students to learn or to bypass the process of learning when doing your assignments.

Glossary

Active Learning - teaching method that actively engages students in the learning process through activities and discussions, as opposed to passively receiving information from lectures.

API (Application Programming Interface) - set of rules and protocols for building and interacting with software applications, which enables different software systems to communicate. Used by developers for example to develop advanced customized chatbots leveraging ChatGPT's technology.

Artificial Intelligence (AI) - field of computer science dedicated to creating systems capable of performing tasks that would typically require human intelligence, such as decision-making and speech recognition.

Backward Design - method of designing educational curriculum by setting goals before choosing instructional methods and forms of assessment.

Chatbot - software application designed to simulate conversations with human users.

ChatGPT - variant of the GPT (Generative Pretrained Transformer) AI models developed by OpenAI, designed specifically to generate human-like text in dialogue.

Custom GPT - version of the ChatGPT model that paid users can tailor or fine-tune for specific tasks or industries to better meet their requirements.

Deep Learning - subset of machine learning involving neural networks with many layers, allowing the model to learn complex patterns in large amounts of data.

Fine-Tuning - process in machine learning where a pre-trained model is further trained (tuned) on a smaller, specific dataset to specialize its knowledge.

Generative AI - AI technologies that can generate text, images, or other media outputs based on training from large datasets.

GPT (Generative Pretrained Transformer) - LLMs developed by OpenAI that use deep learning to produce human-like text by predicting successive words in a sentence.

Hallucination - phenomenon in AI where a model generates incorrect or nonsensical information, not grounded in reality, often due to gaps in training data.

Large Language Model (LLM) - AI model that uses deep learning techniques to understand and generate human language, trained on large amounts of text data.

Machine Learning (ML) - branch of AI that focuses on the development of algorithms that allow computers to learn from and make predictions based on data.

Metacognition - awareness and understanding of one's own thought processes, often used in educational contexts to help students learn how to learn more effectively.

Natural Language Processing (NLP) - branch of AI focused on enabling computers to understand, interpret, and generate human language in a way that is both meaningful and useful.

OpenAI - AI research organization responsible for the development of multiple LLMs, including ChatGPT.

Pedagogical Guardrails - guidelines or measures implemented to ensure that the technology supports effective teaching practices and learning outcomes. In the context of ChatGPT, these are established through fine-tuning a customized chatbot.

Prompt - textual input given to a chatbot (e.g. ChatGPT) to generate a specific output or complete a given task.

Prompt Engineering - practice of crafting prompts to effectively communicate with AI systems, optimizing inputs to achieve the desired output from the model.

Supervised Learning - type of machine learning where the model is trained on labeled data, allowing the algorithm to assess its accuracy and adjust based on performance.

System Prompt - specific instruction or command given to an AI model to guide its performance or behavior, often structured to achieve consistency.

Transfer Learning - pedagogical method where knowledge or skills learned in one context are applied to enhance learning in a new, related area, helping to deepen understanding.

Unsupervised Learning - type of machine learning where the model identifies patterns and relationships in data without any external guidance or labels, enabling it to learn and make inferences from datasets on its own.

Index

About the Authors

Dan Levy has been a faculty member at Harvard University for over 20 years, where he has held various positions related to promoting excellence in teaching and learning. He currently serves as the faculty director of the Public Leadership Credential, the Harvard Kennedy School's flagship online learning initiative. He co-founded Teachly, a web application aimed at helping faculty members to teach more effectively and more inclusively. He has won several teaching awards, including the university-wide David Pickard Award for Teaching and Mentoring. He wrote "Teaching Effectively with Zoom: A practical guide to engage your students and help them learn" and "Maxims for Thinking Analytically: The wisdom of legendary Harvard professor Richard Zeckhauser" His teaching was featured in a book called "Invisible Learning" written by David Franklin. He is passionate about effective teaching and learning, and enjoys sharing his experience and enthusiasm with others.

Angela Pérez Albertos is an MPA in International Development graduate from the Harvard Kennedy School. She is the US Growth Lead at Innovamat, an education organization that focuses on improving math pedagogy across public and private schools and serves as a Teaching Fellow at Harvard University. In the past, Angela has worked as Chief of Staff to Ghanaian education entrepreneur Fred Swaniker, founder of the African Leadership Group, has advised governments on education and job creation policies in Costa Rica, Lebanon, and Rwanda, and has worked as a management consultant at Boston Consulting Group in Madrid. She also holds a Bachelor's degree in Business and International Relations from Universidad Pontificia Comillas (Madrid). Angela is passionate about fostering classroom innovations that motivate and enhance student learning.

344

References and Notes to This Book

[1] "ChatGPT sets record for fastest-growing user base". Reuters (Feb 2023)

[2] "New York City schools ban AI chatbot that writes essays and answers prompts". The Guardian (Jan 2023)

[3] "What is ChatGPT". OpenAI (updated Jun 2024). Available at: https://help.openai.com/en/articles/6783457-what-is-chatgpt

[4] Wei (2022) Chain-of-Thought Prompting Elicits Reasoning in Large Language Models, https://arxiv.org/abs/2201.11903

[5] See for example: Mollick, E., & Mollick, L. (2023, March). Using AI to implement effective teaching strategies in classrooms: Five strategies, including prompts. *SSRN*. https://ssrn.com/abstract=4391243

[6] Demszky, D., Liu, J., Hill, H. C., Jurafsky, D., & Piech, C. (2023, June). Can automated feedback improve teachers' uptake of student ideas? Evidence from a randomized controlled trial in a large-scale online course. *EdWorkingPapers*.

[7] For references on polling, please see:

- Mazur, E. (1997). *Peer instruction: A user's manual*. Prentice Hall.

- Bruff, D. (2009). *Teaching with classroom response systems: Creating active learning environments.* Jossey-Bass.
- Levy, D. (2021). *Teaching effectively with Zoom: A practical guide to engage your students and help them learn.*

[8] As an example, the second part of Derek Bruff's Intentional Teaching podcast episode #41 has a very interesting discussion with his guests Stacey Johnson, Emily Donahoe, and Lance Eaton.

[9] "How ChatGPT Can Help With Grading" by Bruce Ellis (Mar 2023). Available at: https://blog.tcea.org/chatgpt-grading/

[10] Willingham, D. T. (2003). Ask the cognitive scientist: Students remember... what they think about. *American Educator*, 16, 77-81.

[11] Liu, D. (2024). Menus, not traffic lights: A different way to think about AI and assessments. *Teaching@Sydney*. Available at: https://educational-innovation.sydney.edu.au/teaching@sydney/menus-not-traffic-lights-a-different-way-to-think-about-ai-and-assessments/

[12] Mollick, E. (2024). *Co-Intelligence: Living and Working with AI.*

[13] Bowen, J. A., & Watson, C. E. (2024). *Teaching with AI: A Practical Guide to a New Era of Human Learning.*

[14] Shaw, C., et al. (2023). *GenAI in higher education: Fall 2023 update.* Tyton Partners. Available at: https://tytonpartners.com/time-for-class-2023/GenAI-Update

[15] Mollick, E. R., & Mollick, L. (2024). *Assigning AI: Seven Approaches for Students, with Prompts.* The Wharton School Research Paper. Last revised: 14 Jun 2024

Note: Mollick and Mollick suggest seven student uses of Generative AI. Based on our experience and conversations with students, we have adapted the framework from seven to six uses, eliminating the use described as "students: receiving explanations".

[16] Klitgaard, R. (2024). Using ChatGPT in graduate education: A beginner's guide (and we're all beginners here).

[17] Klopfer, E., Reich, J., Abelson, H., & Breazeal, C. (2024). Generative AI and K-12 Education: An MIT Perspective. *An MIT Exploration of Generative AI.*

[18] "How to Create Custom AI Chatbots That Enrich Your Classroom: 4 Examples to Get You Started". Harvard Business Publishing Education (May 2024)

[19] "Introducing the GPT Store". OpenAI (2024). Available at: https://openai.com/index/introducing-the-gpt-store/

[20] "Teaching CS50 with AI - David J. Malan". Harvard CS50 (May 2024). Available at: https://www.youtube.com/watch?v=ggshaJcOc6Y

[21] "Teaching CS50 with AI - David J. Malan". Harvard CS50 (May 2024). Available at: https://www.youtube.com/watch?v=ggshaJcOc6Y

[22] Perplexity AI. Accessed on Jun 22, 2024 at: www.perplexity.ai

[23] Consensus. Accessed on Jun 22, 2024 at: https://consensus.app/